Network and Application Security

Fundamentals and Practices

NETWORK AND APPLICATION SECURITY

Fundamentals and Practices

Debashis Ganguly
Systems Engineer, Infosys Ltd., India

edited by
Shibamouli Lahiri
Department of Computer Science and Engineering,
The Pennsylvania State University,
University Park, USA

CRC Press
Taylor & Francis Group
an **informa** business
www.crcpress.com

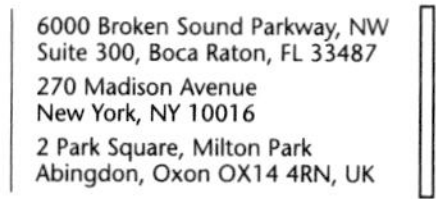

Published by Science Publishers, an imprint of Edenbridge Ltd.
- St. Helier, Jersey, British Channel Islands
- P.O. Box 699, Enfield, NH 03748, USA

E-mail: info@scipub.net Website: www.scipub.net

Marketed and distributed by:

ISBN: 978-1-57808-755-6

CIP data will be provided on request

Printed in the United States of America

To my Maa and Baba,
Ms. Jyotsna Ganguly and Mr. Malay Ganguly,
who mean life to me

Preface

This book is structured into two broad sections that deal with two distinct genres. The following discussion is helpful in understanding the plan behind content alignment and scope of the book.

Part One deals with the basics of network security, threats involved and security measures meant for computer networks.

Chapter 1 gives a general overview on network security and related threats. It helps us get an overall idea on how to deal with security measures.

Chapter 2 talks about standard cryptographic algorithms and their roles in addressing the risks involved in a network as a whole.

Chapter 3 is mainly dedicated to applications like firewalls and IDPS, which provide system-level security.

Chapter 4 illustrates the basics of different applications and standard procedures like Kerberos, X.509 Certificates, PGP, IPSec Suite, SSL, etc., and how they are used to protect the network.

Part Two is meant for discussions related to application-level threats and best coding practices for reducing vulnerabilities that are inherited in applications. This is in contrast with Part 1, which deals solely with network-related issues and discussions. Part 2 gives us a practical outlook on dealing with applications and application-specific network level security.

Chapter 5 talks about several different threats aimed at web applications and web databases. It also helps the reader understand the basic remediation procedures to deal with such threats.

Chapter 6 focuses on .Net and Java coding practices and security guidelines. This chapter helps developers and administrators in gaining a better insight into application-level weak points.

Chapter 7 includes two case studies—one on SAN and the other on VoIP—as security controls for modern day application-specific networks.

Acknowledgements

First and foremost, I would like to acknowledge the painstaking efforts of the hawk-eyed editor Mr. Shibamouli Lahiri, a graduate student in Computer Science and Engineering at The Pennsylvania State University. From adding syntactic sugar to beefing up the pace and style of the book, Shibamouli left no stones unturned to make sure the book proves to be a smooth and enjoyable read for neophytes and experienced users alike. It is my sincere delight that I have found such a good friend and such an impartial reviewer in him.

My earnest acknowledgement goes to the selfless effort put forth by my friend and colleague, Ms. Bhaswati Bhoopalika Das, who stood beside me from the very beginning. Starting from the days when I was struggling with the contents and how to align them to fit into a book, she had helped me with unwavering zeal, be it in accessing more information regarding the content, be it in conducting market surveys to better understand the needs for such a book or be it in reshaping it to get a more practical and effective look.

Last but in no way the least, my sincere gratitude goes to my family, my parents, my teachers and my mentor for all their physical and mental support and an infinite endurance with which they put up with me during this arduous phase of my life. It is really hard to paraphrase my acknowledgement to them, as it will invariably look very little, compared to their so many contributions in my life.

Debashis Ganguly

Contents

Preface vii

Acknowledgements ix

Part One **1**
Network Security: Fundamentals and Practices

Chapter 1: Network Security Fundamentals **3**

1.1 Security Triangle (Three Fundamental Objectives of Network Security) 3
- 1.1.1 Confidentiality 4
- 1.1.2 Integrity 4
- 1.1.3 Availability 5

1.2 Security Threats 5
- 1.2.1 Classification of Network Threats 5
- 1.2.2 Confidentiality Attack 7
- 1.2.3 Integrity Attack 8
- 1.2.4 Availability Attack 9

1.3 Understanding Security Measures 10

Chapter 2: Cryptography and Network Security **15**

2.1 Confidentiality with Symmetric Key Cryptography 15
- 2.1.1 Data Encryption Standard 17

2.1.2 Triple DES 18
2.1.3 Advanced Encryption Standard 20
2.1.4 Key Distribution and Confidentiality 20
2.2 Public Key Cryptography and Message Authentication 22
2.2.1 Overview 22
2.2.2 RSA Public-key Encryption Algorithm 25
2.2.3 Diffie-Hellman Key Exchange 26
2.2.4 Elliptic Curve Architecture and Cryptography 27
2.2.5 Key Management 29

Chapter 3: System-level Security **31**

3.1 Firewall 32
3.1.1 Design Goals behind Firewall 32
3.1.2 Security Controls in Firewall 32
3.1.3 Design Limitations of Firewall 33
3.1.4 Firewall Types 34
3.1.5 Firewall Configuration 39
3.2 Intrusion Detection and Intrusion Prevention Systems 47
3.2.1 Overview 47
3.2.2 Intrusion Detection Systems 48
3.2.3 Intrusion Prevention System 49

Chapter 4: Applications for Network Security **51**

4.1 Kerberos—an Authentication Protocol 51
4.1.1 Overview 51
4.1.2 Implementation Mechanism 52
4.1.3 Analysis 55
4.2 X.509 Authentication Service 56
4.3 Electronic Mail Security 58
4.3.1 Overview 58
4.3.2 Pretty Good Privacy as a Solution to E-mail Security 59

4.4 IP Security 61
4.4.1 Overview 61
4.4.2 Understanding the IPSec Architecture 62
4.4.3 IPSec Implementation 64
4.4.4 Security Association 67
4.4.5 Authentication Header 68
4.4.6 Encapsulating Security Payload (ESP) 70
4.4.7 IPSec Operation Modes 70
4.4.8 Key Management 75
4.5 Web Security 76
4.5.1 Overview 76
4.5.2 Web Security Threats 77
4.5.3 Overview of Security Threat Modelling and General Counter-measures 79
4.5.4 Secure Socket Layer and Transport Layer Security 80

Part Two **91**
Application Security—Fundamentals and Practices

Chapter 5: Application Level Attacks **93**

5.1 Occurrences 93
5.2 Consequences 94
5.3 Attack Types 95
5.4 SQL Injection 95
5.4.1 Overview 95
5.4.2 Consequences 97
5.4.3 Remediation 98
5.5 Cross-Site Scripting (XSS) 101
5.5.1 Overview 101
5.5.2 Consequences 103
5.5.3 Remediation 103
5.6 XML-related Attacks 104

5.6.1 XML Entity Attacks 104
5.6.2 XML Injection 106
5.6.3 XPATH Injection 107
5.6.4 Remediation 107
5.7 Log Injection 108
5.7.1 Overview 108
5.7.2 Consequences 109
5.7.3 Remediation 109
5.8 Path Manipulation 109
5.8.1 Overview 109
5.8.2 Consequences 110
5.8.3 Remediation 110
5.9 HTTP Response Splitting 110
5.9.1 Overview 110
5.9.2 Consequences 110
5.9.3 Remediation 111
5.10 LDAP Injection 111
5.10.1 Overview 111
5.10.2 Consequences 111
5.10.3 Remediation 111
5.11 Command Injection 112
5.11.1 Overview 112
5.11.2 Consequences 112
5.11.3 Remediation 112
5.12 Buffer Overflow 112
5.12.1 Overview 112
5.12.2 Consequences 113
5.12.3 Remediation 113
5.13 Cross Site Request Forgery (CSRF) 113
5.13.1 Overview 113
5.13.2 Consequences 113
5.13.3 Remediation 114

Chapter 6: Practical Software Security—ASP.Net and Java 115

6.1 ASP.Net Security Guidelines 115
6.1.1 Overview 115
6.1.2 Code Access Security (CAS) 116
6.1.3 Windows CardSpace 117
6.1.4 MachineKey Configuration 118
6.1.5 Authentication in .Net 119
6.1.6 Restricting Configuration Override 125
6.2 Java Security Guidelines 126
6.2.1 Java Security Model 126
6.2.2 Specifying Security Constraints 127

Chapter 7: Securing Some Application—Specific Networks 133

7.1 Securing Storage Area Networks 133
7.1.1 Overview 133
7.1.2 Purpose behind SAN 134
7.1.3 SAN Design Components 135
7.1.4 SAN Security Issues 135
7.1.5 Security Measures for SAN 138
7.2 Securing VOIP-enabled Networks 139
7.2.1 Overview 139
7.2.2 Why VoIP? 139
7.2.3 VoIP Design Components 141
7.2.4 VoIP Security Issues 141
7.2.5 Security Measures for VoIP 143

Index 147

Part One

Network Security: Fundamentals and Practices

CHAPTER 1

Network Security Fundamentals

Just as *absolute* does not exist, a completely secure network is not possible. With the growth of the internet and telecommunication, newer techniques to breach the security of a network have evolved. It is important to keep in minds that in the scenario of network vulnerability, even an unsophisticated and innocent-looking user can pose a serious threat to an unprotected network. However unintended the attack may be, its consequences can be severe and substantial. Attackers, equipped with knowledge of the latest technological advances, pose threats that may bypass detection mechanisms of a network and consequences of such attacks include financial loss and loss of trust. This chapter deals with the primary goals of network security, different flavours of network vulnerabilities and understanding of counter-measures.

1.1 SECURITY TRIANGLE (Three Fundamental Objectives of Network Security)

In the last few decades, the IT world has witnessed a huge growth in internet and e-commerce applications. With more demanding customers, the companies are obliged to introduce more data centres and more network devices for ensuring improved turnaround time and higher website availability. As a result, networks continue to grow in size and become more complex. Managing and protecting them from security threats therefore assume paramount importance, as it increasingly becomes more tedious and painful to handle such chores manually. Today's network administrators are thus required to understand the three fundamental objectives of network security which constitute the security triangle.

1.1.1 Confidentiality

Confidentiality means the preservation of data privacy from unwanted and illegal users. The secrecy of data can be achieved both logically and physically by restricting access to information in its most expressive, raw, legible form. It can be achieved by the following means:

1. Physically seclude sensitive data from the reach of network users.
2. Use logical barriers like Access Control Lists (ACL) or firewalls to protect it from an invalid access.
3. Credentials-based authentication process to get through the gateway of the network.
4. Network traffic encryption, so that even if an attacker infiltrates the data, it cannot be deciphered.

1.1.2 Integrity

Data Integrity refers to the preservation of content and source of the data. It consists of checking whether the data has been transmitted from an authentic source and has not been tampered with in transit. The following measures can help achieve Data Integrity:

1. Parity checking for detecting data modification,
2. Public key cryptography and digital signatures for validating the authenticity of the origin of the data.

1.1.3 Availability

Data Availability refers to the duration in which user data can be made available on the network. It is nearly synonymous with system uptime. Higher Data Availability can be ensured by the following means:

1. Preventive measures for protecting application networks from Denial of Service attack.
2. Proper data filtering to minimize the processing of junk data.

1.2 SECURITY THREATS

An overwhelming majority of modern day network attacks target the application layer rather than the lower layers. One major reason behind this shift of focus is that modern attackers are mostly profit-oriented, as opposed to amateur hackers looking for satisfaction gained from breaking into someone's highly secured network. It therefore becomes very important for today's network administrators to have solid knowledge on several possible security threats, their nature and possible remedies. This section first explains the categorization of threats and then cites a few major vulnerabilities of networks and network-based applications.

1.2.1 Classification of Network Threats

The moment you get connected to the World Wide Web, you become vulnerable to threats from outsiders who can intervene and steal information. But it should not be construed to mean that a network completely cut-off from the World Wide Web is totally safe either, because internal users can still pose a threat to it. Depending on the source, threats can be classified as:

- **Internal Threats:** This kind of threats is more serious than the external ones, because
  - o insiders already possess legitimate access to some portions of the network

- o they have knowledge about the network structure, security mechanism and weaknesses in-built into the system and
- o traditional security schemes like Intrusion Prevention Systems and firewalls are ineffective, since internal users work from within their protection boundary.

- **External Threats:** External attackers do not have proper knowledge about the network beforehand. So, before attack they try to gather more information about the network by monitoring traffic, call walking, etc. The next step is to identify a single flaw in the whole mechanism by Brute Force Attack. Once the attacker is sufficiently acquainted with the network and its properties, (s)he may launch even more sophisticated and focused attacks to further cripple the system. This kind of attacks is more technical and challenging in nature.

Network security threats can further be classified into the following categories based on position, attitude and purpose of the attacker:

- **Passive:** A passive attacker does not harm the logistics of the network; (s)he only passively snoops the information and leaks it to the outside world.
- **Active:** In an active attack, the attacker disrupts the functionality of a network by injecting a lot of junk data into it. It usually destroys the data in transit, and the attack is fairly easy to identify from the mangled data.
- **Close-in:** As its name suggests, a close-in attacker remains in close proximity of the target system and gains physical access to it.
- **Insider:** An Insider Attack occurs when legitimate users of the system try to obtain unauthorized privileges by overriding their assigned access rights.
- **Distributed:** In a Distributed Attack, the attacker creates some loopholes or access points in the system that is hidden from legal users. Whenever a legal user steps over such a loophole, a piece of information silently goes to the attacker. In this way,

the attacker gradually gains control over the whole target system and in turn the entire network.

1.2.2 Confidentiality Attack

It is a passive form of attack where the attacker attempts to obtain confidential information about network users like login credentials, SSN, Credit Card information or e-mail password. This kind of attack may go undetected if the attacker masquerades as a legitimate user and then snoops private information, rather than trying to tamper with the data or crash the system.

In most cases, application server, web server and database server interact with each other based on mutual trust relationships. So when an attacker becomes able to compromise the confidentiality of a web server, eventually (s)he gets access to the sensitive data stored in database server as well.

An attacker can launch a confidentiality attack in the following ways:

1. Dumpster Diving: The attacker obtains credentials and other private information from un-shredded papers dumped in office bins.
2. Social Engineering: In most applications, users tend to generate passwords based on their dates of birth, some family-member's name, etc. An attacker can socialize with the target user to obtain his/her personal details, and then use that information for guessing passwords.
3. Wire Tapping: If the attacker is located in close physical vicinity of the target network, then (s)he can tap into network lines and snoop over secret messages.
4. Packet Capture: The attacker can easily capture data packets travelling across a network. Therefore, by systematically intercepting a hub with which the victim is connected, or by tricking the packets to flow through his system by acting as a honeypot, the attacker can obtain a lot of sensitive information.

5. Ping Sweep and Port Scanning: An attacker can flood a network with a list of pings and capture positive responses from one or some of the pings. This allows him/her to glean a list of IP addresses of all network devices. After successfully locating device IP addresses, the attacker scans a range of servicing UDP and TCP ports to identify potential targets.

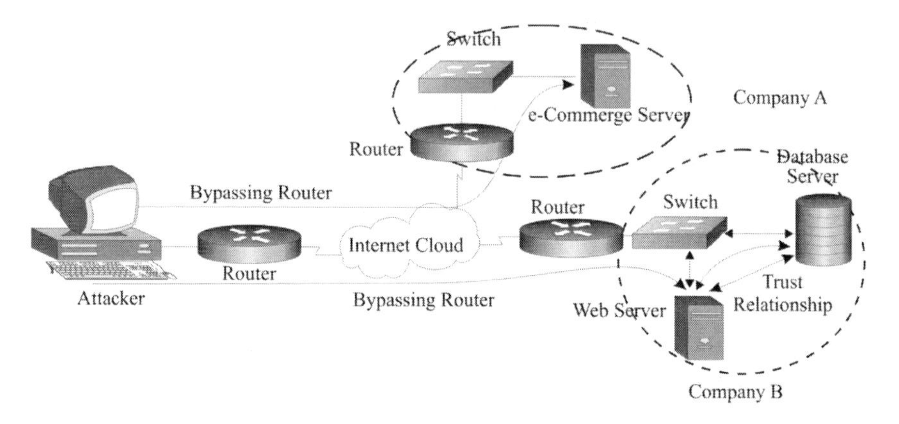

Figure 1.1: Understanding Confidentiality Attack.

1.2.3 Integrity Attack

Integrity attack is based on confidentiality attack, except that the attacker does not stop after snooping data; rather, (s)he tries to *modify* the content.

An integrity attack can be launched in the following ways:

1. Botnet attacks: The attacker writes a piece of software called network robot ("botnet" in short) and injects it into the target system. This malicious piece of code makes the whole infected system act like a slave, thereby compromising the integrity and confidentiality of huge amounts of data.
2. Password attacks using Trojan horse, packet capture, key logger application or dictionary attacks to obtain user credentials from the system.
3. Hijacking legitimate TCP sessions.

4. Salami attack: Salami attack or Penny Shaving attack is a series of smaller attacks, which taken together engenders a devastating consequence.

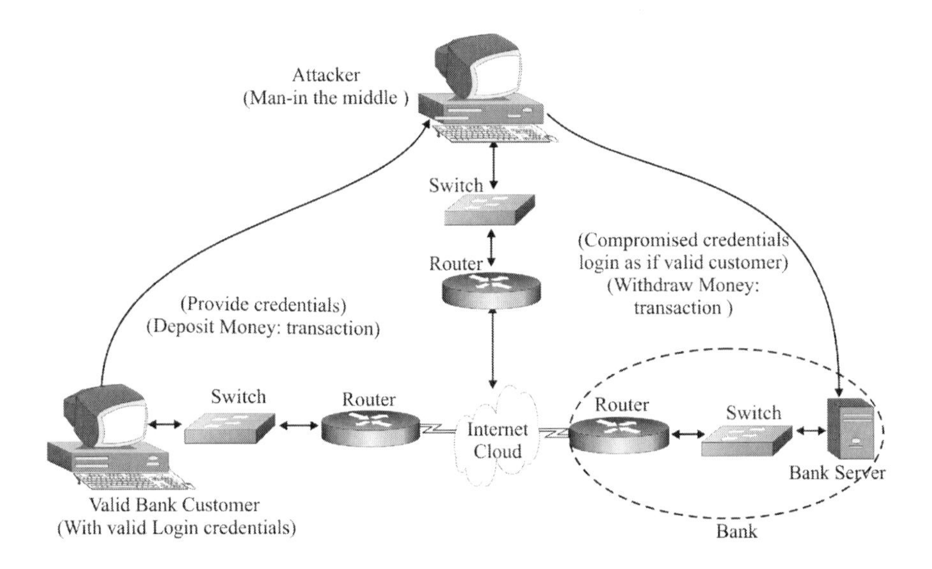

Figure 1.2: Understanding Integrity Attack.

1.2.4 Availability Attack

Availability attacks attempt to limit the usability and accessibility of network resources by tricking the system to process a huge amount of unwanted payload. It can also be affected by the crude means of damaging devices physically.

A few varieties of availability attacks are listed below:

1) **DoS and DDoS:** The attacker floods one or many of the systems in the victim network with junk data and forces those systems to process the data. It causes the victim network to exhaust its CPU cycles and memory slots, thereby failing to provide service to the legitimate users. When the flooding comes from a single source, it is known as a Denial of Service (DoS) attack; when it comes from multiple sources in different subnets, it is called a Distributed Denial of Service (DDoS) attack.

2) **TCP SYN Flood:** This is a variant of DoS, where the attacker floods a network with one-way TCP SYN messages, but never sends back ACK signals in response to those SYN messages. This leads to an avalanche of incomplete, unattended three-way handshakes for TCP session establishment. As many of the servers limit the number of acceptable TCP sessions, they fail to process valid handshake requests after hitting the limit, and henceforth become unavailable.
3) **ICMP Attack:** Much like the TCP SYN flood, here the attacker floods a target network with ICMP pings causing DoS. In this variant, the attacker masquerade as a troubleshooter of the victim network and starts sending copious amounts of ICMP pings, as if for diagnostics purpose. These ICMP pings are always incomplete and intentional, so the system soon gets overwhelmed and a DoS situation is reached.

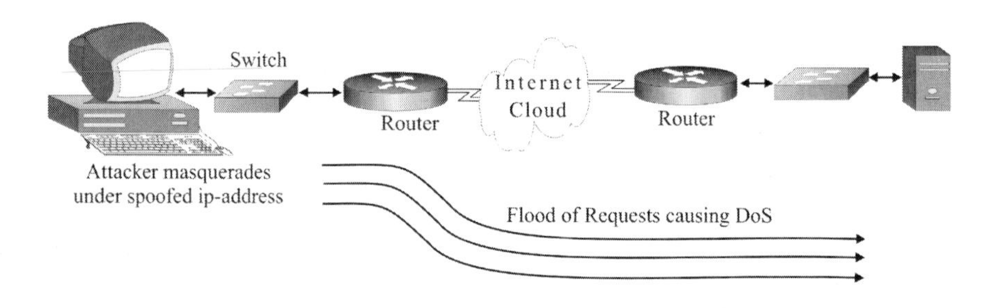

Figure 1.3: Understanding Availability Attack.

1.3 UNDERSTANDING SECURITY MEASURES

Before securing data in the network, it is important to understand relative sensitivity of the data and classify it in a hierarchy of access levels. Not all data is of the same importance. Neither do all data need the same level of security. In an application domain, a network administrator ought to categorize data according to its level of sensitivity and then apply the appropriate security mechanisms to it. The following are some of the criteria for categorizing data according to its relative sensitivity:

- **Value:** How valuable is the data to the organization?
- **Age:** How old is the data?
- **Relevant Life Span:** How long does the network administrator think the data is going to be relevant?
- **Personal Touch:** How much personal information does the data contain?

A successful categorization of data allows users to be classified into roles, as follows:

- **Owner:** People who own the data. The owners are responsible for determining the classification hierarchy, the level of security applicable to the data, and frequency of data reviews for identifying faults.
- **Custodian:** These are the people who are responsible for securing the data, keeping it updated, managing faults, and planning and executing backups.
- **User:** These are normal users who usually have a limited access to one particular view of the data, within the purview of security policies.

With data classification and role classification put into place, administrators should have a clear understanding of the risk potential. Understanding the vulnerabilities of a system can best be done by donning the hat of an attacker. In particular, we need to assess the following:

- **Motive:** What can possibly drive the attacker to launch a particular attack?
- **Means:** With all security controls in place, how can an attacker carry out the attack?
- **Opportunity:** Can the attacker have an opportunity to break into the system, given that (s)he is available to launch an attack?

A thorough understanding of the above-mentioned questions will help a network administrator look for controls in a security solution. Controls in a security solution can be broadly classified as follows:

- Administrative controls: These are primarily policy-centric, e.g.,
  - o Routine security awareness programs.
  - o Clearly defined security policies.
  - o A change management system.
  - o Logging configuration changes.
  - o Properly screening potential employees.
- Physical controls: These help protect the data's physical environment and prevent potential attackers from easily having physical access to the data.
  - o Intrusion Detection Systems (IDS).
  - o Physical security barriers and security personnel to safeguard the data.
  - o Biometric identification of visitors.
- Technical controls: These are a variety of hardware and software technologies for protecting data.
  - o Security applications like firewalls, IPSs, and VPN termination devices.
  - o Authorization applications like one-time passwords (OTP), Single Sign-On (SSO) and biometric security scanners.

The administrative, physical, and technical controls can further be classified as one of the following control types:

- o Preventive: A preventive control attempts to restrict access to data or a system.
- o Deterrent: A deterrent control attempts to prevent a security incident by de-motivating the potential attacker and thereby keeping him/her from the process of launching an attack.
- o Detective: A detective control is meant to detect an illegal or invalid access to the data or a system.

Recommended Readings and Web References

CippGuide: CIA triad page.

CsoOnline: Page on the magic triangle of IT security by Michael Oberlaender.

Pfleeger, Charles P. 1997. Security in Computing; Upper Saddle River, NJ; Prentice Hall.

Watkins, Michael and Kevin Wallace. 2008. CCNA Security Official Exam Certification Guide; Cisco Press.

CHAPTER 2

Cryptography and Network Security

Cryptography is an art of hiding a message from unintended listeners. It was practised at the dawn of civilizations just like it is practised today, albeit in a different persona. Today, cryptography is used to provide secrecy in the world of computers and the internet. In the scenario of network vulnerability, cryptography can be used to satisfy three major aspects of security—Confidentiality, Integrity and Authenticity. The primary focus of this chapter is to discuss several types of cryptographic protocols that help ensure network security.

2.1 CONFIDENTIALITY WITH SYMMETRIC KEY CRYPTOGRAPHY

Symmetric key cryptography is the standard way of maintaining confidentiality where a single, secret key is used to both encrypt and decrypt data. It is thus also referred to as "secret-key" or "single-key" cryptography. A symmetric encryption scheme is based on the following functional components:

- Algorithms: A cryptographic algorithm is a set of actions designed to produce encrypted data from plain text ("encryption algorithm"), or plain text from encrypted data ("decryption algorithm").

- Plain text: A text message composed by the end-user. It is input to an encryption algorithm and output from a decryption algorithm.
- Secret Key: Usually a long string of bits. Along with plain text, the secret key is also input to an encryption algorithm. Note that the decryption algorithm also needs this secret key to produce the correct plaintext output.
- Cipher text: A scrambled message produced as the output of an encryption algorithm and sent across a network to the intended recipient. The recipient usually has to perform a decryption of the cipher text to get back the original message.

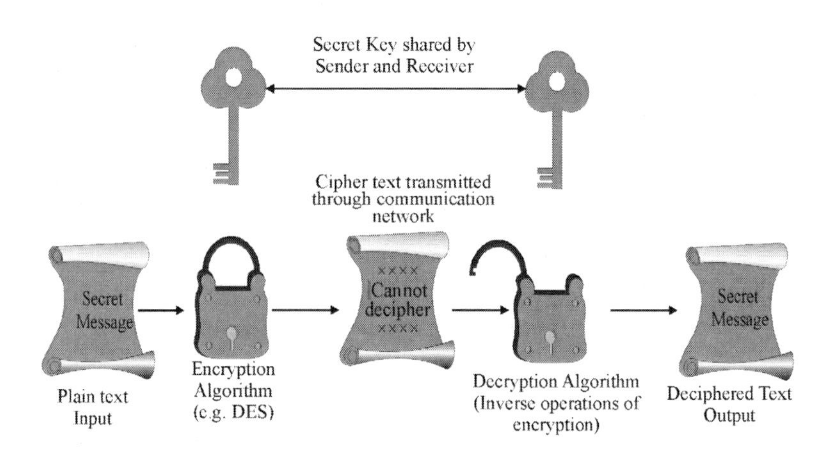

Figure 2.1: Confidentiality Using Symmetric Key Cryptography.

The most commonly used symmetric encryption algorithms are block ciphers. A block cipher processes the plaintext input in fixed-size blocks and produces a block of cipher text of equal size. Three standard algorithms in symmetric key cryptography are the Data Encryption Standard (DES), the Triple DES (3DES), and the Advanced Encryption Standard (AES). A brief overview on these approaches, their shortcomings and their benefits is given below.

2.1.1 Data Encryption Standard

2.1.1.1 Overview

The Data Encryption Algorithm (DEA), better known as the Data Encryption Standard (DES), is a symmetric key encryption algorithm that belongs to the family of block ciphers. Being a block cipher, it works on a block of data. The block size is 64 bits, i.e., 8 bytes. The secret key, which is used both in encryption as well as in decryption, has an overall size of 64 bits, i.e., 8 bytes, but only 56 bits (7 bytes) are used.

In DES, the first step is to generate 16 sub-keys from the 56-bit secret key. Each data block is XOR-ed 16 times with each of these generated sub-keys in a criss-crossing manner. Being a symmetric key protocol, decryption uses the same secret key as for encryption, but in the decryption algorithm, sub-keys are used in the reverse order of encryption. DES employs the Feistel Network model of encryption with minor modifications.

2.1.1.2 Modes of operation in DES

Like any other block cipher, DES can be run in two major modes—ECB (Electronic Codebook) and CBC (Cipher Block Chaining). In the ECB mode, each 64-bit chunk of the input message is treated independently. ECB is therefore very vulnerable to brute-force attacks, because all you need to decode is any one of those 64-bit chunks, and the rest becomes easy. In the CBC mode on the other hand, the encryption algorithm is initially fed with a 64-bit seed block. This seed block is used to encrypt the first 64-bit message block. Then the resulting cipher text is used to encrypt the second 64-bit message block, which in turn encrypts the third 64-bit message block, and so on.

2.1.1.3 Speed of DES

DES decomposes the secret key into 16 sub-keys and then iteratively operates on the 64-bit data blocks. These operations make DES slow. Hence, it is best suited for encryption of moderate to small amounts of data. If the data is large and yields a large number of 64-bit blocks, DES becomes very slow and should be avoided for all practical purposes. Storing data in string blocks rather than arrays helps DES perform much faster.

2.1.1.4 Need to find an algorithm superior to DES

The strength of DES, like any other symmetric key algorithm, lies principally on the choice of the key. It has been observed over time that the DES algorithm itself is pretty much impenetrable, since it decomposes the secret key into 16 sub-keys over 56-bit key space and repeats the encryption procedure in a Feistel scheme. But the issue of key choice remains, and has given rise to serious implications. It was initially presumed that the effective key size being 56 bits, a potential attacker has to guess the key from a space of 2^{56} ($\approx 7.2 \times 10^{16}$) keys, making it almost impossible for him/her to finish the brute-force process within a reasonable amount of time with reasonable amount of resources. But this presumption was later proved to be false by the Electronic Frontier Foundation (EFF). They designed a dedicated chipset that broke a DES encryption within 3 days. This called into question the effectiveness of DES as a strong encryption algorithm and necessitated the development of another algorithm with higher strength and larger key space.

2.1.2 Triple DES

2.1.2.1 Overview

The Triple Data Encryption Algorithm (TDEA) is a block cipher commonly known as Triple DES (3DES). 3DES applies DES thrice to each data block. 3DES was introduced to give better protection towards brute-force attack than DES, although it could not completely replace DES or the applications that continued relying on it. Triple DES was designed to provide a relatively simpler encryption solution after the weakness of DES as an encryption standard became public. While 3DES uses the same core logic as DES, it comes with a larger effective key size—168, 112 and 56 bits, respectively—the last one being present for backward compatibility with DES. The key sizes can be specified as *keying options* 1, 2 or 3, respectively (keying option is an input parameter to the 3DES algorithm). 3DES is much less vulnerable to the types of brute force attacks that spelled the end of DES.

2.1.2.2 Encryption and decryption algorithms

Triple DES uses a bundle of three DES keys—K_1, K_2 and K_3—each of 56 bits (excluding parity bits).

The encryption algorithm is the following:

CT : E_{K3} (D_{K2} (E_{K1} (PT))) where,
CT : Cipher text,
PT : Plain text,
E_{Ki} : DES encryption with K_i
D_{Ki} : DES decryption with K_i

The novelty of 3DES lies in the fact that each block of input message goes through three independent DES operations; a DES encryption with K_1, then a DES decryption with K_2 and finally another DES encryption with K_3.

Decryption follows the reverse procedure:

$$PT = D_{K1} (E_{K2} (D_{K3} (CT)))$$

Having a combination of three distinct DES operations vastly improves the strength of the algorithm. Note, however, that the maximum strength can *only* be achieved under keying option 1, as discussed in the next section. Keying option 2 provides moderate strength, whereas option 3 is meant for backward compatibility with DES.

2.1.2.3 *Keying options*

The standards define three keying options:

- Keying option 1: All three keys are independent. This is the strongest option, because the effective key size is 3 × 56 = 168 bits and the effective key space is 2^{168} ($\approx 3.74 \times 10^{50}$) keys.
- Keying option 2: K_1 and K_2 are independent, while K_3 is the same as K_1. This provides moderate security, with 2 × 56 = 112 effective key bits. But this option is stronger than DES—encrypting the plaintext twice.
- Keying option 3: All three keys are identical, i.e. $K_1 = K_2 = K_3$: This is equivalent to DES, with only 56 effective key bits. Option 3 provides backward compatibility with DES.

2.1.3 Advanced Encryption Standard

The Advanced Encryption Standard (AES) is a symmetric-key encryption standard introduced as a replacement for its predecessors—3DES and DES. The AES consists of three block ciphers—AES-128, AES-192 and AES-256, where each of these ciphers has a 128-bit block size and key sizes of 128, 192 and 256 bits respectively. The AES ciphers have undergone extensive cryptanalysis and are now used worldwide. A noteworthy exception in AES compared to DES and 3DES is that AES does not follow the Feistel Network structure for block cipher encryption. Instead, it performs one permutation and three substitution steps, carefully customized to enhance the strength of the algorithm.

2.1.4 Key Distribution and Confidentiality

Considering a standard algorithm generally used in applications, the overall security of the symmetric encryption scheme finally boils down to the strength and secrecy of the key. Although in a certain sense it makes all symmetric encryption schemes look somewhat vulnerable, the principal appeal of these schemes lies in their standardizations which makes widespread use and low-cost chip-level deployment feasible. As you can probably guess by now, the principal security problem in symmetric key encryption schemes is maintaining the secrecy of the key. Therefore, secure and efficient *key distribution* plays a vital role in symmetric key cryptography.

For two communicating parties A and B, there are several key distribution options:

1. Party A can select a key and *physically deliver* it to B.
2. A third party can select a key and deliver it to A and B.
3. If A and B have previously communicated with a key, then they can use the previous key to *encrypt a new key* and send it across the network.
4. If A and B have a secure encrypted communication channel with a third party C, then C can relay key between A and B.

Physical key delivery is the simplest, but only possible if both parties have personal contact with each other. In the case of a large number of parties communicating with each other, physical key delivery is virtually impossible for two reasons—the parties may not know each other, and physical delivery does not scale well. This lack of scalability is what necessitated the introduction of Session Key Management and Key Distribution Centres (KDC).

Provided that an appropriate algorithm has been chosen and a secret key acquired by both communicating parties, it can be ensured that the message sent from the sender is encrypted in a way that its content is completely obscured to any unintended recipient; and that upon receiving the message at the recipient end, only a legal user can decrypt it with the same shared key. Thus, symmetric key encryption provides confidentiality of a secret message across a potentially unreliable network.

2.2 PUBLIC KEY CRYPTOGRAPHY AND MESSAGE AUTHENTICATION

2.2.1 Overview

Although symmetric key cryptography is widely in use for providing network security and message confidentiality, an urge was felt to look for some alternative cryptographic algorithm types. The main reason behind this quest was that the entire strength of symmetric key algorithms effectively lay in their use of secret keys. If somehow the secret key is compromised, the whole process of encryption and decryption becomes meaningless. And usually there is no foolproof scheme available for sharing a secret key across an unsecured network. This big shortcoming paved the way of *public key cryptography*. The first scheme in public key cryptography was introduced by Diffie and Hellman.

In contrast with the simple operations on bit patterns commonly employed by the secret key system, the public key scheme usually relies on relatively complex mathematical functions. Unlike the former, it uses two keys—the *Public Key* and the *Private Key*. Thus, it has another name—Asymmetric Key Cryptography. It is important to note that Public-key cryptography may not provide more security than Private-key cryptography, as the strength of the public-key mechanism often depends on the complexity of the algorithm and the length of the keys. Also note that the public key cryptography was not introduced to replace its predecessor; rather, it is most likely the case that both types of cryptosystems will continue to co-exist owing to the mathematical complexity of public key system.

Message Authentication is a major security aspect where public key cryptography has enjoyed widespread use. The symmetric key cryptosystem completely fails to address the requirements of message authentication, whereas the public-key cryptography not only addresses those requirements, but is also able to provide confidentiality.

• *Message confidentiality*

Provision of confidentiality through public key algorithms can be summarized into the following points:

- Consider two communicating parties A and B. Each of them generates a pair of Public keys KU_a, KU_b and Private keys KR_a, KR_b. They keep their respective private keys to them as the name suggests and publish the respective public keys. Thus, each user gets to know the public keys of all other users.
- Sender A, knowing the public key of recipient B and having the message to be encrypted M, generates the cipher text:

$$CE_{KUb}(M)$$

- A then sends the encrypted text to B. B decrypts the cipher text with its own private key and recovers the original message:

$$M = D_{KRb}(C) = D_{KRb}[E_{KUb}(M)]$$

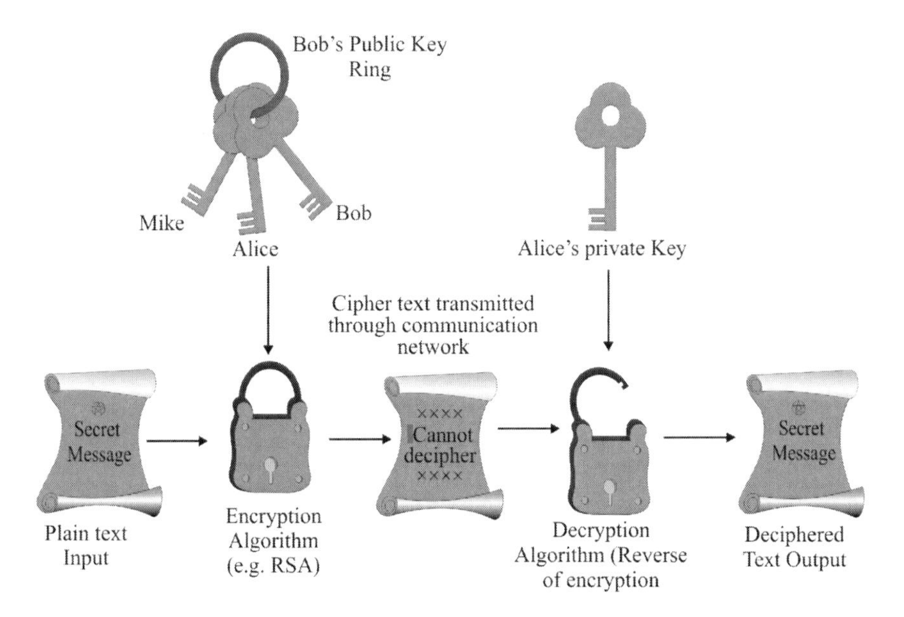

Figure 2.2: Message Confidentiality Using Public Key Cryptography.

- It is computationally infeasible or very hard for an opponent (who knows the public key KU_b) to determine the private key KR_b, and thereby decrypt the cipher text.
- Either of the two related keys can be used for encryption, with the other used for decryption.

$$M = D_{KRb}[E_{KUb}(M)] = D_{Kub}[E_{KRb}(M)]$$

• *Message authentication and data integrity*

Message Authentication can be achieved in exactly the same way as Data Confidentiality, except that the order of usage of the two keys is reversed. In authentication, the sender does not worry about the importance or secrecy of the message. (S)he only encrypts the message with his/her Private Key and then sends it along to the intended recipient. Usually the message contains User Identification data which, the sender hopes, will prove his/her authenticity to the intended recipient. The sender-generated ciphertext is known as the Digital Signature of the sender. When recipient receives this digital signature, (s)he

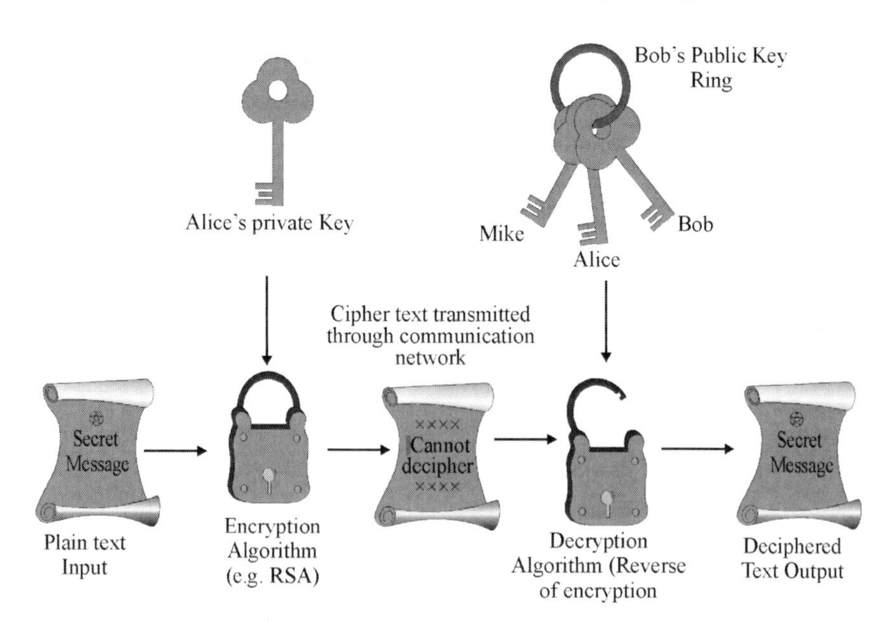

Figure 2.3: Message Authentication Using Public Key Cryptography.

attempts to decrypt it with the sender's public key. On success, it is verified that the digital signature came from the sender and nobody else, because no other person than the sender could possibly be in possession of the former's private key. Thus, authentication of a user is achieved.

As the message is encrypted using a private key, no one can alter its content in transit. So, tamper resistance and thus data integrity is ensured.

2.2.2 RSA Public-key Encryption Algorithm

RSA is one of the most popular public key algorithms. It was developed as a standard and is widely used in current systems. Introduction of RSA has revolutionized the fields of cryptography and network security in many ways.

RSA uses a block cipher in which the plaintext and the cipher text are blocks of integers between 0 and n-1 for some integer n. Given a plaintext block M and a cipher text block C, encryption and decryption are of the following form:

$$C = M^e \bmod n$$

$$M = C^d \bmod n = (M^e)^d \bmod n = M^{ed} \bmod n$$

where e and d are special integers, as described later in this section. Both sender and receiver must know the values of n and e, and only the receiver knows the value of d. So, the public key KU = {e, n} and the private key KR = {d, n}.

RSA provides satisfactory security performance given that the following assumptions hold:

1. It is possible to find values of e, d, n such that $M^{ed} = M \bmod n$, for all $M < n$.
2. It is relatively easy to calculate M^e and C^d, for all values of $M < n$.
3. It is infeasible or extremely expensive to determine d given e and n.

The key generation of RSA is difficult as it involves a great deal of complex mathematics. The steps in key generation can be summarized as follows:

1. Two distinct prime numbers p and q of the same bit length are chosen at random and n is computed as the product of p and q, i.e., $n = pq$.
2. $\Phi(n) = (p - 1)(q - 1)$ is computed, where φ is Euler's Totient function.
3. Then, an integer e is calculated such that $1 < e < \varphi(n)$ and e and $\varphi(n)$ are co-prime.
4. Another integer d is calculated as the multiplicative inverse of e (modulo $\varphi(n)$).

2.2.3 Diffie-Hellman Key Exchange

In 1976, Diffie and Hellman proposed a key agreement protocol (also known as "exponential key agreement") in their paper ("New Directions in Cryptography") that enabled two users to exchange a secret key across an insecure medium without any prior communication.

The protocol revolves around two parameters p and g which are both public and shared by all users in a system. The parameters p and g are chosen such that:

- p is a prime number
- g (usually called a generator) is an integer, $g < p$, and
- for every number n, $1 \leq n \leq p-1$, there exists k such that $n = g^k \bmod p$.

The Diffie-Hellman protocol is as follows. First, the two communicating parties (say A and B) derive their own private values a and b randomly from the set of all integers. Then using the public values p and g, they compute their public keys ($g^a \bmod p$) and ($g^b \bmod p$) respectively, and share them with each other. Then each party computes $g^{ab} = (g^b)^a$, and $g^{ba} = (g^a)^b$ (all modulo p). Since $g^{ab} = g^{ba} = k$ (all modulo p), finally A and B both have a shared secret key k.

The protocol depends on the discrete logarithm problem for its security. The discrete logarithm problem ensures that it is computationally infeasible or extremely expensive to calculate the shared secret key $k = g^{ab} \bmod p$ given the two public values $g^a \bmod p$ and $g^b \bmod p$ when the prime p is sufficiently large.

2.2.4 Elliptic Curve Architecture and Cryptography

As discussed in the previous sections, the strength of any cryptographic algorithm, whether public key or private key, largely depends on two factors: mathematical or operational complexity of the algorithm and the length of the keys. Although RSA as a public-key cryptosystem is widely in use for message authentication while private key systems are used for ensuring confidentiality, they all are limited by the time and space complexity of the algorithms and by the size of keys that needs to be ensured.

Thus, an alternative scheme was introduced for public-key cryptography based on the algebraic structure of elliptic curves on finite Galois Fields. The Elliptic Curve architecture emerges from a harder and more complex mathematical domain than those containing the discrete logarithm (DL) or integer factorization (IF) problems. But the beauty of elliptic curve algorithms lies in their lower space and time complexity, as well as substantially smaller key size. Thus Elliptic Curve Cryptography is slowly being adopted by mobile devices and applications like Cellular phones, Pagers, PDAs, etc.

Although the concept of Elliptic Curves is by no means a new one, it did not see much use or implementation so far because of the mathematical difficulties involved. Naturally there is a lot of scope for advancement and further innovations. As the system is yet to be standardized for a general platform, there are lots of implementations available in the market based on adaptations of different curve specifications.

The concept of Elliptic Curve Architecture can be understood as follows:

Consider a finite Galois Field GF (p), $p > 3$, and let $a, b \in$ GF (p) are constants such that $4a^3 + 27b^2 \equiv 0 \pmod{p}$.

An elliptic curve, $E_{(a,\, b)}$ (GF (p)), is defined as the set of points $(x, y) \in GF(p) * GF(p)$ that satisfies the equation $y^2 \equiv x^3 + ax + b \pmod{p}$ together with a special point, O, called the *point at infinity*.

Let P and Q be two points on $E_{(a, b)}$ (GF (p)) and O the point at infinity. From the properties of modulo addition, we obtain:

- P + O = O + P = P
- If P = (x_1, y_1) then –P = $(x_1, -y_1)$ and P + (–P) = O
- If P = (x_1, y_1) and Q = (x_2, y_2), and P and Q are not O, then P + Q = (x_3, y_3), where
  - o $x_3 = \lambda^2 - x_1 - x_2$
  - o $y_3 = \lambda(x_1 - x_3) - y_1$ and
  - o $\lambda = (y_2 - y_1)/(x_2 - x_1)$, if $P \neq Q$
  - o $\lambda = (3x_1^2 + a)/ 2y_1$, if P = Q

Based on the formulation above, the elliptic curve domain parameters over GF (q) are defined as a sextuple: T = (q, a, b, G, n, h), where

- $q = p$ or $q = 2^m$
- a and $b \in GF(q)$
  - o $y^2 \equiv x^3 + ax + b \pmod p$ for $q = p > 3$
  - o $y^2 + xy = x^3 + ax^2 + b$ for $q = 2m \geq 1$
- a base point G = (x_G, y_G) on $E_{(a, b)}$ (GF(q)),
- a prime $n \equiv$ the order of G. The order of a point P on an elliptic curve is the smallest positive integer r such that rP = P + P + P + ... (r times) = O.
- h = #E/n, where #E is the number of points on the elliptic curve and is called *curve order.*

A public key Q = (x_Q, y_Q) for a communicating entity associated with a sextuple (q, a, b, G, n, h) is generated using the following steps:

- A random or pseudo-random integer d is selected such that $1 \leq d \leq n-1$.
- Compute Q = dG = G + G + G + ... (d times).
- Q becomes the public key and d the private key of the communicating entity.

2.2.5 Key Management

Key management in cryptography refers to the generation, exchange, storage, safeguarding, use, vetting, and replacement of keys. Key management is important in cryptographic protocol design and it includes considerations for key servers, user procedures, types of key management that needs to be adopted, and other relevant protocols.

Successful key management is the most critical and arguably the most difficult aspect of a cryptographic system. Note that the symmetric key cryptosystem is somewhat limited in being able to address the risks involved in key management. The public-key cryptosystems on the other hand, have taken it as a challenge. One of the major roles of public-key encryption is to address the problem of key distribution. Two distinct uses of public-key encryption in key management are as follows:

1. Distribution of public keys

In public key system, each communicating party generates his/her own pair of private and public keys. Whereas the private key remains privy to a single user, the public key is broadcast to all other users. Broadcasting public keys is not always effective and secured. For example, an attacker can masquerade as a valid party and send a spoofed public key to all users before the legitimate party gets a chance. In such cases, if other users send a message to the legitimate party, it goes to the attacker instead. Thus, newer technologies have evolved to securely publish and share public keys between communicating users and systems. Technologies like X.509 issue public key certificates for individual users. The authenticity of a user is established up front by a secret sharing of information between certificate-issuing authorities (CA) and users. After a certificate has been obtained, all users subscribed to a particular CA can communicate and share the certified public keys among each other.

2. Distribution of secret keys

Sharing secret keys for symmetric key cryptography is one of the challenging aspects addressed by the public key cryptosystem. Diffie-Hellman Key Exchange protocol and public key certificates or Kerberos can be used in this regard.

Recommended Readings and Web References

csrc.nist.gov: AES Algorithm (Rijndael) Information.

csrc.nist.gov: Recommended Elliptic Curves For Federal Government Use.

DeviceForge.com: An Elliptic Curve Cryptography (ECC) Primer.

Koblitz, Neal. 1987. Elliptic curve cryptosystems, in Mathematics of Computation 48, pp. 203–209.

Miller, Victor S. 1985. Use of elliptic curves in cryptography, CRYPTO 85.

Recommended Elliptic Curve Domain Parameters, Certicom Research, Certicom Corp., Version 1.0, September 20, 2000.

Rivest, Ron, Adi Shamir, Leonard Adleman. A Method for Obtaining Digital Signatures and Public-Key Cryptosystems; Communications of the ACM. 21(2):120–126.

RSA.com: RSA Laboratories.

Salomaa, Arto. 1996. Public Key Cryptography, New York: Springer-Verlag.

Stallings, William. 2003. Cryptography and Network Security: Principles and Practices; 3rd edition; Upper Saddle River, NJ: Prentice Hall.

Chapter 3

System-level Security

With the introduction of computers and the internet, the world witnessed a rapid change in the mode of business. Currently, almost all organizations—ranging from the government to small private enterprises—have either largely computerized their state of affairs or are in the process of doing so. Thus, modern offices have become synonymous with sets of workstations connected to each other and sharing information in real time. With the number of processing demands from customers running an all-time high, these networks have become quite complex in nature—ranging from LANs and Edge Networks to WANs. Note, however, that the organization employees often need to connect to the internet for running their daily chores, and the organizations themselves also need to connect to the internet to have an online presence. But the moment the internet is connected to the internal network, the latter becomes subject to a wide range of vulnerabilities and network security threats from external users. To protect the internal network, the servers and the organizational data from such external threats, a new application known as the "firewall" was introduced. Like its counterpart in building construction, a firewall protects the internal network from the "fire" of external threats coming from the internet. Along with firewalls, intrusion detection and prevention systems also play a key role in deterring or removing external attacks. This chapter presents a first-hand overview to such systems.

3.1 FIREWALL

3.1.1 Design Goals behind Firewall

A Firewall is a mechanism for controlling the flow of data between two communicating networks, e.g., between the internet and an intranet. Followings are the design goals for coming up with a good firewall:

- All traffic from the inside network to the outside network, and vice versa, must pass through the firewall.
- Only authorized traffic adhering to the norms and policies defined in the firewall will be allowed to pass.
- The firewall itself will be immune to penetration. In other words, the firewall must not break down under traffic pressure, or under malicious traffic.

3.1.2 Security Controls in Firewall

A firewall acts as the first line of defence for the internal network. Its roles are as follows:

- An observer, which monitors all activity from outside or from inside the network,
- An analyzer, which analyzes the traffic according to a set of pre-defined policies, and A restrictor, which prevents all unwanted and unauthorized access.

A firewall plays all the above roles by enforcing some basic controls on the network and the network users' behaviour. The controls are as follows:

- Service Control: Rules that specify which categories of services are valid for the network.
- Direction Control: Rules that specify whether the allowed services are allowed inbound or outbound or both ways.

- User Control: Rules that specify users' access and roles for being eligible to use one or more of the services and direction controls combination.
- Behaviour Control: This is the most complex type of control to implement in practice. It refers to a set of rules that tries to determine how the other three controls should work together. It defines the modes and ways of operation for all combinations of the other three controls.

3.1.3 Design Limitations of Firewall

A firewall is generally able to deliver the following basic security requirements:

- It acts as a single point of access between the external and the internal networks and thus can serve the purpose of a security check point.
- It can be used to monitor security-related issues originating from the external or the internal networks.

But there are certain drawbacks and limitations that a firewall is often subject to. These are as follows:

- A firewall is difficult to design, because it is often really problematic in practice to come up with an effective set of policies for the firewall. A good firewall also costs a lot of money.
- Since a firewall acts as a "choke point" between the internal and the external networks, it can severely limit the external network access. For example, a conservative set of security policies may allow only a few designated websites to be browsable by internal users, while restricting all others. Thus, a firewall can excessively limit the internet usage, often without a good enough justification for doing so.
- A firewall is meant to protect internal data from external users with the assumption that a network itself is safe from the internal users. This assumption, however, may often prove to be wrong. Internal users often pose a serious threat to the internal network, which is at least as great as or

sometimes even greater than the external threats. Firewalls are completely useless in the face of such internal threats.

- Some firewalls, e.g., the Packet Filtering Router check the source address of IP packets as a first step to filtering the traffic. But if an attacker spoofs a valid source address and sends packets with modified headers containing the spoofed IP address, then the firewall will consider these packets valid and will allow them to pass through. Thus, firewalls are of no use in the face of such masquerading attacks.
- A firewall in its basic form cannot give protection from malicious codes, virus infected programs and files, Trojan horses or spam mails.
- If an attacker is able to fragment packets into very small chunks and repeat the TCP header, then (s)he may be able to bypass a firewall as well. This is possible only if the firewall has been configured to check the first packet and allow the rest when the first one is correct, and the attacker knows this.
- A firewall cannot encrypt e-mail messages or private documents going out of the internal network.
- Even if there is a firewall in place between WAN and the internet as a gateway and gives protection against external threats, each individual user can connect to the internet using a dial-up connection from their respective workstations. The organizational firewall cannot help such individuals.

3.1.4 Firewall Types

A firewall can either be a software bundle coming with an existing operating system or it can be a piece of hardware. The hardware firewall can either be a single system or a set of two or more systems working together to perform the designated functionality. Depending on the mode of filtering and operability, firewalls can broadly be classified into following categories:

• *Packet filtering router*

Network layer firewalls or packet filters operate at a lower level of protocol stack, i.e., the network layer. It applies a set of rules to all inbound and outbound IP packets and filters out packets that do not follow one or more of the rules. Depending on how it was configured, a firewall may either allow almost all packets; rejecting only a few or it may reject almost all packets while allowing only a few. An IP packet can be filtered based on the following information stored in the packet header:

- Source IP Address,
- Destination IP Address,
- Source and destination transport layer address (i.e., port address),
- IP protocol field,
- Router Interface
- IP ports.

A packet filtering router can be classified into two categories:

- **Stateless Firewalls**

  These are the firewalls that filter packets mainly based on network layer information and do not require transport or session layer context. These filters need less memory and can work faster as they do not need to investigate the *session* associated with a particular connection. Thus they limit themselves from performing more complex decision making based on which stage of communication has been reached between hosts.

- **Stateful Inspection Firewalls**

  Unlike the stateless firewalls, stateful firewalls record session information and use that information to process packets. Since most client/server applications run on top of TCP protocol, it is relatively easy to record session information for those applications, as TCP sessions can be succinctly characterized by a small set of variables like source and destination IP addresses, port information and session lifetime.

  If a new packet does not match the context of an existing session (determined by the firewall's state table), then the firewall inspects it for the beginning of a new session. On the other hand, if the packet does match an existing session, then it is allowed to pass through sans any further processing. It means when a new packet comes, then this type of firewall validates the based on rule and takes decision accordingly; but if the packet is already a part of some previous connection, it will allow the packet without further investigation.

Although packet filtering routers are appealing because of their simplicity and speed, they do have limitations. In particular, they fail to give protection from

- o Attacks targeting the higher levels of protocol stack (i.e., application layer), and
- o Masquerading attacks (e.g., IP spoofing)

• *Application-level gateway*

Unlike packet filters, application-level gateways operate at the application layer. They act as *proxy servers*, sitting between workstations and the actual server that connects to the external world. Their unique position gives them a special advantage. They can detect and discard malicious packets that are pouring in from the external world, before they get to reach the internal traffic. Workstations (i.e., end-users) generally communicate with the gateway via Telnet or FTP (which are all TCP/IP applications) and the gateway asks them for names or IP addresses of the remote hosts they want to access. Upon receipt of this information, the gateway contacts one or more applications located on the remote host and then relays TCP segments containing application data to the workstations, given that the segments are valid and non-malicious.

This kind of firewalls patches the pitfalls of packet filtering routers by operating at a higher level of the protocol stack. Instead of checking packet headers, it checks application information and the traffic content to come to a reject/accept decision. Thus, it has the potential to prevent threats from virus, malicious codes, suspicious mails and corrupt files. Application-level gateways can be further configured to allow only some specific features of a particular application which administrators found acceptable and either block or disable all other features.

With such flexibilities and advantages over packet filtering routers, application-level gateways are widely used in organizations. Note, however, that they implement a host of complex logic that are often responsible for slowing down network traffic due to additional processing load.

• *Circuit-level gateway*

A circuit-level gateway can either be a sophisticated, dedicated stand-alone system or it can be a specialized function performed by an application-level gateway. It acts as a bridge between two TCP hosts (an internal host and an external peer) and hence establishes two TCP connections. This prevents a direct connection between the internal and the external hosts, thereby removing the possibility of a potentially unsecured communication in the system. By acting as a hop point and relaying TCP segments from one host to the other, a circuit-level gateway validates connection authenticity before allowing its establishment.

A typical scenario of operation is when it operates on the basic assumption that the internal users mean no harm to the network and thus is generally configured to support proxy services for inbound traffic and circuit level functionality for outbound datagram. Thus it reduces complexity of validating outbound traffic as it inspects only the session state and operation before establishment of connection. It Circuit-level gateways typically operate at the session layer of ISO model or at the network layer of TCP/IP suite.

• *Personal firewall*

Personal firewalls are meant for the general public—individual users using a dial-up or a broadband connection to connect to the internet. Note that the organizational firewalls, the ones we talked about so far, are usually not optimized to work at the personal level. Hence the attacker can easily compromise the security of individual workstation and thus with this victim system can gain control over whole internal network by spreading botnets, virus, worms, Trojan horses. In such scenarios, personal firewalls act as a barrier between an individual user and the internet, thereby protecting the former from botnets, virus, worms, Trojan horses, etc. Even when a user is not connected to office LAN and is only working on his home computer, it is advised to incorporate personal firewall to get protection against internet based threats.

A personal firewall can either be

- o A software installed in individual computers which works in tandem with the operating system, or it can be
- o A piece of hardware installed in the router or the modem.

The software firewalls are cheap and often come free-of-cost with existing operating systems. They are useful not only for personal use, but also for small office setups.

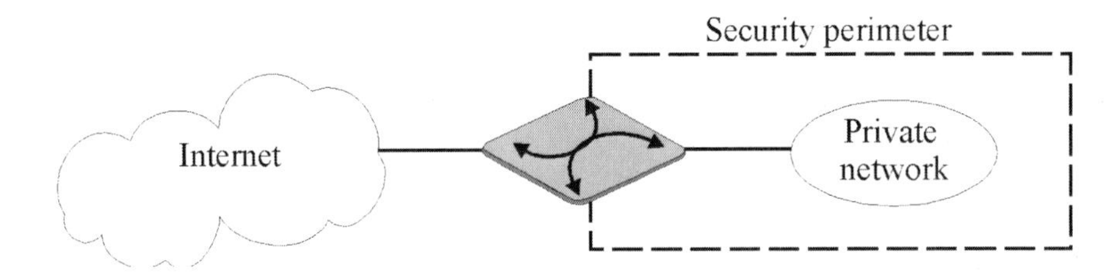

Figure 3.1: Packet Filtering Route.

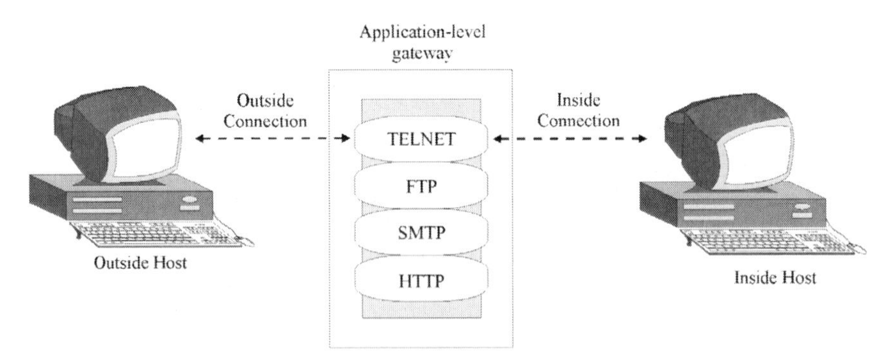

Figure 3.2: Application Level Gateway.

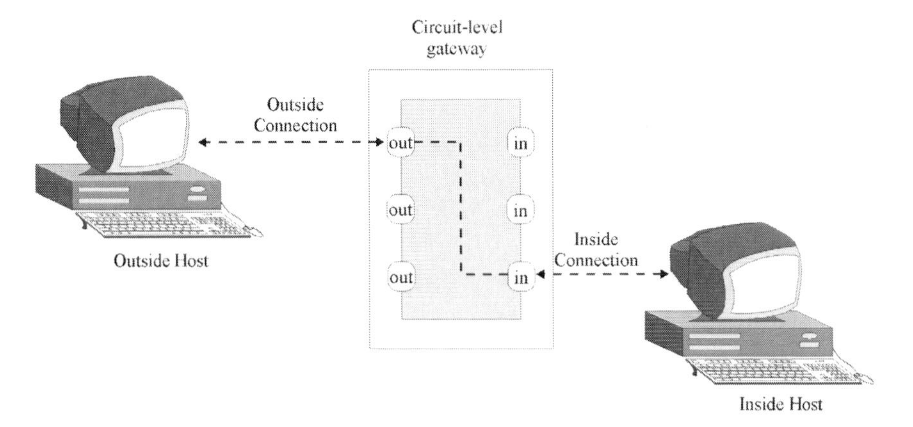

Figure 3.3: Circuit Level Gateway.

3.1.5 Firewall Configuration

There are many firewall solutions available in the market that can be further customized and configured to fit the business requirements of an organization. Configuration of an organizational firewall involves making decisions about the way the set of devices should communicate with each other and the policies regarding those communications. Most organizational products available in the market use a dedicated hardware unit that can be customized by defining a set of rules for filtering traffic. The rule list is actually a set of appropriately sequenced system commands that helps define the criteria for filtering. Coming up with an acceptable rule list for the whole organization is usually very difficult, as it requires understanding of all security needs of the organization, effective business optimization of those needs, and translating them to a set of rules.

One widely-used organizational security solution is the Cisco ASA 5500 Series Adaptive Security Appliances line. They are hardware filtering units designed to support organizational security by playing the roles of a Network Address Translator (NAT), an Intrusion Prevention System and a Virtual Private Network (VPN). ASA is more a switching platform than a router; hence it also provides hardware acceleration.

Although the Cisco ASA has an undeniable appeal in being a singleton device that performs a host of security services, it is usually not advisable to force a single unit to do everything, especially when traffic loads are high. In such cases, filtering tasks can be split up across several different units. For example, software firewalls can be employed in parallel with hardware firewalls to speed up processing. Cisco Internetwork Operating System is one such software that operates on network switches and routers. It works as a packet filter and uses its own Access Control List (ACL). The ACL is a rule list, i.e., a set of system commands standardized as part of IOS. The ACL has several variations and versions. The Standard ACLs and the Extended ACLs are most popular. A standard ACL filters packets based on source IP address only, whereas an Extended ACL looks for both source as well as destination IP addresses. Extended ACLs can further be configured to take into account protocol type, port numbers, sequence numbers, precedence values, etc.

The syntax for Standard ACLs is as follows:

```
access-list access-list-number {permit|deny}
{host|source source-wildcard|any}
```

The following example defines a standard access control list for a firewall setup between two networks NetA and NetB through gateway R1:

- **Allow a Select Host to Access the Network**

  All hosts in NetB are denied access to NetA, except a single host with IP address 192.168.10.1.

R1

```
hostname R1
!
interface ethernet0
ip access-group 1 in
!
access-list 1 permit host
192.168.10.1
```

- **Deny a Select Host to Access the Network**

  All hosts in NetB are allowed access to NetA, except the host with IP address 192.168.10.1. Note that since there is an implicit "deny all" condition in an ACL, the sequencing of commands is very critical in successfully configuring it.

R1
hostname R1 ! interface ethernet0 ip access-group 1 in ! access-list 1 deny host 192.168.10.1 access-list 1 permit any

- **Allow Access to a Range of Contiguous IP Addresses**

  In the following example, 0.0.0.255 is the inverse mask of the network 192.168.10.0 with mask 255.255.255.0. ACLs use the inverse mask to select a range of contiguous IP addresses. In this example, the ACL permits communication between all hosts with source addresses in the 192.168.10.0/24 network and all hosts with destination addresses in the 192.168.200.0/24 network.

R1
hostname R1 ! interface ethernet0 ip access-group 101 in ! access-list 101 permit ip 192.168.10.0 0.0.0.255 192.168.200.0 0.0.0.255

- **Deny Telnet Traffic (TCP, Port 23)**

  Depending on security needs, it sometimes becomes necessary to block some specific applications, e.g., Telnet.

The following configuration demonstrates how to deny the Telnet application, but allow all others from NetA to NetB.

```
R1
hostname R1
!
interface ethernet0
ip access-group 102 in
!
access-list 102 deny tcp any any
eq 23
access-list 102 permit ip any any
```

- **Allow Only Internal Networks to Initiate a TCP Session**

  The ACL in this example allows hosts in NetA to initiate and establish a TCP session to hosts in NetB and denies hosts in NetB from initiating and establishing TCP sessions to hosts in NetA. This example relies on the fact that all well-known TCP ports are between 0 and 1023. Thus, any datagram with a destination TCP port less than 1023 is denied by ACL 102.

```
R1
hostname R1
!
interface ethernet0
ip access-group 102 in
!
access-list 102 permit tcp any
any gt 1023 established
```

- **Deny FTP Traffic (TCP, Ports 20 and 21)**

  The following ACL configuration shows that FTP control (TCP, port 21) and FTP data (port 20) traffic sourced from NetB destined to NetA are denied, while all other traffic is permitted.

R1

```
hostname R1
!
interface ethernet0
ip access-group 102 in
!
access-list 102 deny tcp any any
eq ftp
access-list 102 deny tcp any any
eq ftp-data
access-list 102 permit ip any any
```

- **Allow FTP Traffic (Active Mode)**

  FTP can operate in two different modes—active and passive. In active mode, the FTP server uses port 21 for control and port 20 for data. With the assumption that the FTP server (192.168.1.100) is located in NetA, the following ACL shows that FTP control (TCP, port 21) and FTP data (TCP, port 20) traffic sourced from NetB destined to FTP server (192.168.1.100) is permitted, while all other traffic is denied.

R1

```
hostname R1
!
interface ethernet0
ip access-group 102 in
!
access-list 102 permit tcp any host
192.168.1.100 eq ftp
access-list 102 permit tcp any host
192.168.1.100 eq ftp-data established
!
interface ethernet1
ip access-group 110 in
!
access-list 110 permit host
192.168.1.100 eq ftp any established
access-list 110 permit host
192.168.1.100 eq ftp-data any
```

- **Allow FTP Traffic (Passive Mode)**

  In passive mode, the FTP server uses port 21 for control and the dynamic ports greater than or equal to 1024 for data. Considering the FTP server (192.168.1.100) located in NetA, the next ACL configuration shows that FTP control (TCP, port 21) and FTP data (ports greater than or equal to 1024) traffic sourced from NetB destined to FTP server (192.168.1.100) is permitted, while all other traffic is denied.

R1
hostname R1 ! interface ethernet0 ip access-group 102 in ! access-list 102 permit tcp any host 192.168.1.100 eq ftp access-list 102 permit tcp any host 192.168.1.100 gt 1024 ! interface ethernet1 ip access-group 110 in ! access-list 110 permit host 192.168.1.100 eq ftp any established access-list 110 permit host 192.168.1.100 gt 1024 any established

- **Allow Pings (ICMP)**

  The following ACL shows that ICMP pings sourced from NetA destined to NetB are permitted, but pings sourced from NetB destined to NetA are denied, thereby eliminating all ICMP packets except an echo-reply.

```
R1

hostname R1
!
interface ethernet0
ip access-group 102 in
!
access-list 102 permit icmp any
any echo-reply
```

- **Allow HTTP, Telnet, Mail, POP3, FTP**

  The following ACL configuration demonstrates that only HTTP, Telnet, Simple Mail Transfer Protocol (SMTP), POP3, and FTP traffic are permitted, and the rest of the traffic sourced from NetB destined to NetA is denied. Services are referred to by names and ports.

```
R1

hostname R1
!
interface ethernet0
ip access-group 102 in
!
access-list 102 permit tcp any
any eq www
access-list 102 permit tcp any
any eq telnet
access-list 102 permit tcp any
any eq smtp
access-list 102 permit tcp any
any eq pop3
access-list 102 permit tcp any
any eq 21
access-list 102 permit tcp any
any eq 20
```

- **Allow DNS**

  The next configuration shows that only Domain Name Service (DNS) traffic is permitted, and the rest of the traffic sourced from NetB destined to NetA is denied.

R1
hostname R1 ! interface ethernet0 ip access-group 102 in ! access-list 112 permit udp any any eq domain access-list 112 permit udp any eq domain any access-list 112 permit tcp any any eq domain access-list 112 permit tcp any eq domain any

Although we limited our discussions to configurations with a single device, in practice we observe more complex configurations having a combination of different devices aligned to work in tandem to give a complete security solution for organizations. In such setups, Bastion Hosts play a key role. A Bastion Host is a fortified node in the network that adversaries cannot easily breach. It can be an application-level or a circuit-level gateway, working in collaboration with other filtering devices and routers. A few types of complex device combinations are described below.

Screened Host Firewall and Single-homed Configuration is a setup where the internal network is connected to a bastion host and a packet filter is used to interface between the internet and the internal network. This setup adds the flexibility of packet-filters to the power of an application-level gateway by allowing all traffic to pass through if they are destined for the bastion host. But if an attacker is able to compromise the network layer firewall, then (s)he can force the traffic flow to bypass the bastion host. To address this problem, the concept of "dual-homed configuration" came into play. In a dual-homed configuration, there are two separate network layer interfaces in the bastion host—one meant for the internet and the other for the internal network. It therefore breaks the direct network layer connection between the internet and the internal network, thereby forcing all traffic to pass through the bastion host.

In another type of complex combination, two separate packet filters are used. One router is placed between the internet and the bastion host and another router is placed between the bastion host and the internal network. This is an alternative to the dual-homed configuration, known as Screened Subnet Firewall Setup. The key advantage of this configuration is that even if an adversary is able to compromise one of the routers, (s)he cannot break into the system because of the other router.

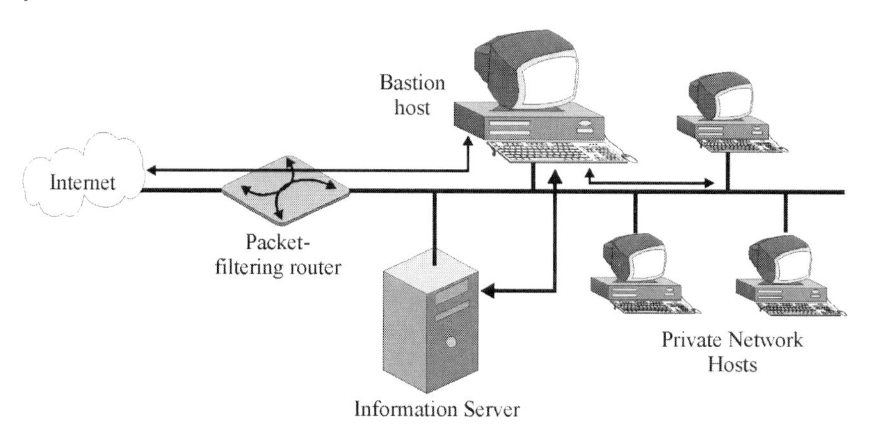

Figure 3.4: Screened host firewall system (Single-homed bastion host).

3.2 INTRUSION DETECTION AND INTRUSION PREVENTION SYSTEMS

3.2.1 Overview

Intrusion Detection Systems and Intrusion Prevention Systems (IDPS) are security measures that work as an extension of a traditional firewall where the latter fails to give adequate protection. A firewall, although effective for external threats, is virtually of no use for internal threats. An IDPS, on the other hand, works in the face of internal attacks as well. Where a firewall generally stops an attack by filtering the traffic, an IDPS monitors the traffic and investigates it for an intrusion. If an intrusion does occur or is suspected to have occurred, an alarm is raised immediately. An IDPS can also tell if an intrusion is currently being planned by monitoring the traffic.

3.2.2 Intrusion Detection Systems

An Intrusion Detection System is a device or software like application-level gateways that monitors network traffic for malicious or suspicious activities and raises alarms to alert network administrators. An IDS can belong to either of the two major families:

• *H-IDS*

Host-based Intrusion Detection Systems (H-IDS) are software like application-level gateways that work with a broad range of operating systems. It is generally installed as a daemon service which monitors application logs, audit trails, file system activities (e.g., changes in ACLs, password files, binaries, etc.) and other host activities, and investigates the types of recent changes to detect a possible threat. An H-IDS serves only one host.

It requires no extra hardware and gives support against threats like virus, worms, and Trojan horses.

• *N-IDS*

Network-based Intrusion Detection Systems (N-IDS) are hardware devices connected to a network hub or a switch that act as choke points. An N-IDS typically utilizes a network adapter running in promiscuous mode to monitor and analyze network traffic in real-time.

An N-IDS is not limited to a single host like an H-IDS. In comparison with H-IDS, it is more difficult for attackers to remove the evidence of attack from N-IDS, since the latter is completely independent of the host environment.

3.2.3 Intrusion Prevention System

Intrusion Prevention Systems (IPS), more commonly known as Intrusion Detection and Prevention Systems (IDPS), are an extension of IDS. They detect network intrusions, raise alarms and then go ahead to prevent the attack. They are very similar to an application-level gateway, and perform four tasks—identify suspected incidents, log information about them, attempt to stop them and report them to system administrators.

The main difference between an IDS and an IPS is that the latter does not passively listen to the traffic like the former; rather, it blocks the traffic immediately if a malicious activity is smelled. An IPS does not depend on environmental factors like the protocol in use, and can stop intrusion by naïve means like dropping the packet bearing offending content or header, and dropping and resetting the connection.

IPSs are available in different varieties (much like IDSs):

- Network-based Intrusion Prevention Systems,
- Wireless Intrusion Prevention Systems,
- Network Behaviour Analyzers,
- Host-based Intrusion Prevention Systems.

IPSs follow three main methods for intrusion detection:

- Signature-based detection,
- Statistical Anomaly-based detection, and
- Stateful Protocol Analysis detection.

Recommended Readings and Web References

CISCO Press: Intrusion Prevention: Signatures and Actions by Earl Carter, Jonathan Hogue.

CISCO Website: Cisco Secure IPS—Excluding False Positive Alarms.

CISCO Website: Configuring Commonly Used IP ACLs.

en.kioskea.net: Introduction to intrusion detection systems.

en.kioskea.net: Intrusion prevention systems (IPS).

Firewall.com: Explore most of the information available.

Zwicky, Elizabeth D., Simon Cooper and D. Brent Chapman. 2000. Building Internet Firewalls, O'Reilly & Associates, June 2000.

CHAPTER 4

Applications for Network Security

This chapter talks about several different applications and tools designed for enforcing network security. Detailed discussions are given on two authentication mechanisms—Kerberos and X.509. In the second part of this chapter we focus on E-mail, IP and Web security.

4.1 KERBEROS—AN AUTHENTICATION PROTOCOL

4.1.1 Overview

Kerberos was designed as part of project Athena at MIT. Unlike other authentication mechanisms, Kerberos relies on symmetric-key encryption. It uses the Needham-Schroeder symmetric protocol as its basis for authentication.

Kerberos assumes an open network where users at their respective workstations try to access services distributed across the network. In such an environment, it is possible for a user to gain access to a particular device or a system by masquerading as another user or to modify device data and make others believe this is due to some other application. Kerberos attempts to address these issues.

The current version of Kerberos is v5. It is more complex than the previous version (v4), and it provides more security rules than v4. Instead of designing authentication mechanism at each server distributed over network, Kerberos provides a centralized authentication server that authenticates user's identity and also validity of a service providing server. Kerberos deals with a scenario, where users in a distributed environment, uses a time-sharing operating system and time synchronized network resources.

This form of centralized authentication acts as a reliable third party for secret key sharing between clients and servers. Here, users and servers have their own private keys, by means of which they authenticate themselves to the centralized server, and once their identities are validated, the central server allows them to interconnect and communicate with each other through secured channel based on time-stamped session key and service tickets.

4.1.2 Implementation Mechanism

Kerberos has the following key structural components:

- A Key Distribution Centre (KDC), which consists of
  - o An Authentication Server (AS) and
  - o A Ticket Granting Server (TGS)
  - Client Workstations,
  - Service Server,
  - Ticket-to-get-ticket (TGT)

The phases involved in Kerberos authentication mechanism are as follows:

- User Client-based Logon:
  1. A user enters a username and a password on the client workstation.
  2. The client system performs a one-way function (hash usually) on the entered password, which becomes the secret key of the user.
- Client Authentication:
  1. The client sends a text message containing the user ID to the AS.

2. The AS searches for the User ID in its database (e.g., Active Directory in case of Windows Server) and generates a secret key by hashing the password of the user found in its database.
3. The AS sends back the following two messages to the client:
   - o Message A: TGS Session Key encrypted using the secret key of the user.
   - o Message B: TGT, containing client ID, client network address, ticket validity period, and the TGS session key, encrypted using the secret key of the TGS.
4. The client tries to decrypt message A with the secret key generated from the user-entered password. If this decryption is successful, the client is able to retrieve the TGS session key. Otherwise, the client gives an "incorrect password" message to the user.

- Client Service Authorization:
  1. When requesting services, the client sends two messages to the TGS:
     - o Message C: Contains the TGT from message B and the requested service ID.
     - o Message D: An Authenticator (i.e., client ID and timestamp), encrypted using the TGS Session Key decrypted from message A.
  2. Upon receiving messages C and D, the TGS retrieves message B out of message C and decrypts the former (with TGS secret key) to get the TGS session key. Using TGS session key, the TGS is able to decrypt message D. Then the TGS sends two messages to the client:
     - o Message E: Client-to-server ticket (composed of client ID, client network address, validity period and Server Session Key), encrypted with the service's secret key.
     - o Message F: Server Session key encrypted with the TGS Session Key.

- Client Service Request:
  1. Upon receiving messages E and F from the TGS, the client sends two messages to the SS to authenticate itself and get access to the service:
     - o Message E from the previous step
     - o Message G: a new Authenticator, (consisting of the client ID and timestamp), encrypted with the Server Session Key.

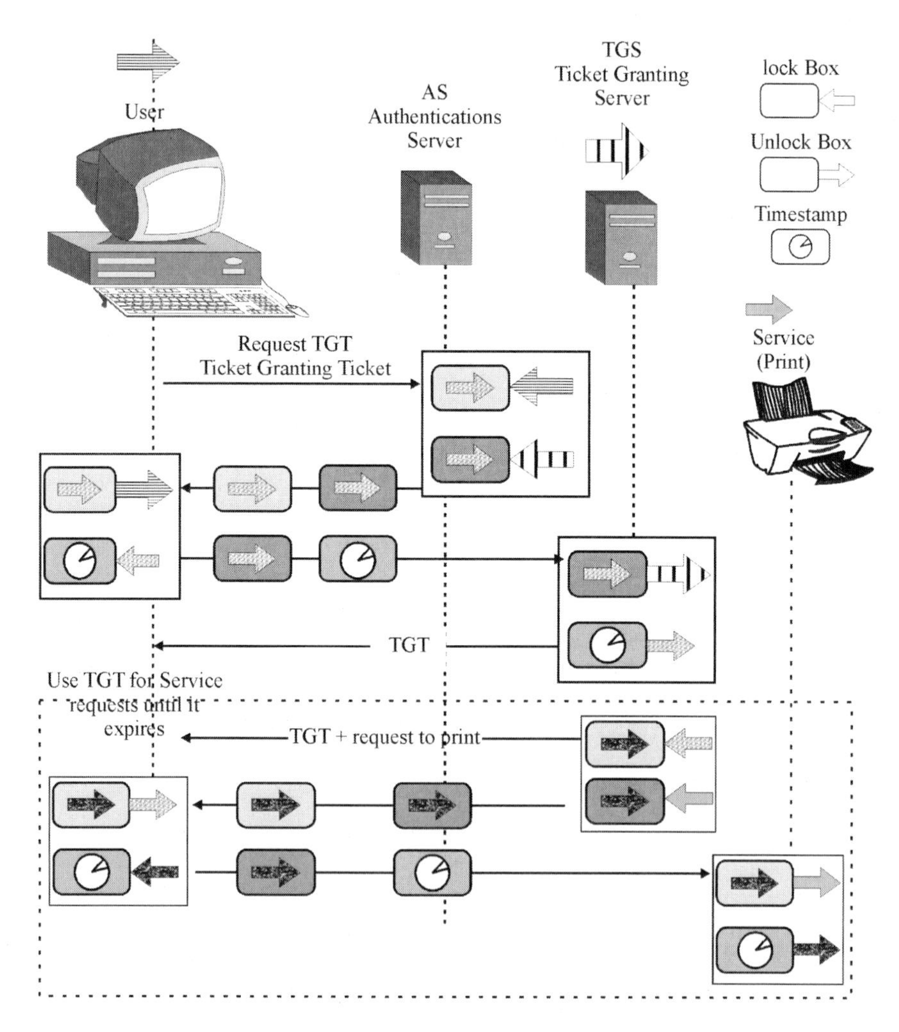

Figure 4.1: Functionality of Kerberos.

2. The SS decrypts message E with its own secret key to retrieve the Server Session Key. Using the Server Session Key, SS decrypts the new Authenticator (message G) and sends a message back to the client to authenticate itself and grant access to the client:
   - o Message H: the timestamp found in message G plus 1, encrypted with the Server Session Key.
3. The client decrypts message H with Server Session Key and checks whether the timestamp has been correctly updated. On success, the client can trust the server and can start issuing further service requests to the server.

4.1.3 Analysis

Although Kerberos v5 is now widely accepted as a standard authentication protocol, it has some important drawbacks:

- In Kerberos, the Central Server (i.e., the KDC) is responsible for individual client and service authentication. So, continuous availability of the KDC is a must. Whenever the KDC is down, no one can log in.
- In Kerberos, clocks of the individual hosts must be synchronized. The protocol requires no clock be more than five minutes apart from each other. The tickets have an availability period and if the host clock is not synchronized with the KDC clock, authentication may fail.
- Since the overall authentication procedure is controlled by the KDC, the latter represents a weak point in the system. If a user is able to compromise the KDC, the whole system suddenly becomes vulnerable. Compromising the KDC is not too difficult either, because all you need to do is masquerade as a system administrator.

These drawbacks can be addressed using the following steps:

- Issues related to the availability of KDCs can be mitigated by employing multiple KDCs operating in tandem. Extra KDCs can be employed for ensuring redundancy and reliability within the system.

- The clock synchronization issue is not so severe in practice, because Network Time Protocol daemons are used to keep the host clocks synchronized with each other.

4.2 X.509 AUTHENTICATION SERVICE

A directory is a specialized server or a distributed set of servers dedicated to maintain user information like the user name, network ID, different user attributes and even the public key certificate of the user. X.500 defines such kinds of directory services. Since we are talking about huge public network systems and millions of users communicating with each other, it becomes necessary for each individual user to verify the authenticity of a public key certificate obtained by another user or in the more extreme case, the public key certificate of a certification authority (CA). X.509 is part of the X.500 series of standard recommendations, which addresses the authentication of public key certificates obtained by network users and signed by a CA. It defines a framework for authentication to be used by the X.500 directories. X.509 is a widely practiced authentication scheme which is in use by several different security mechanisms like SSL, SET, SHTTP, PEM, PKCS, S/MIME, etc.

The latest version of X.509 is v3. It addresses the security issues involved in its predecessors—v1 and v2. X.509 protocol does not mandate the use of any standard algorithms. Thus, users or authentication agents can use any algorithm and still follow the steps specified by X.509. The major revision in v3 is the use of public key cryptography (like RSA) for authentication, whereas v1 and v2 mostly relied on Secret Key cryptography for authentication purposes.

The core of X.509 v3 is standardization of public-key certificates associated with individual users. These certificates are created by trusted CAs (subscribed by users) and stored in their databases. An X.509 certificate consists of the following fields:

1. version
2. serial number

3. signature algorithm ID
4. issuer name
5. validity period
6. subject (user) name
7. subject public key information
8. issuer unique identifier (versions 2 and 3 only)
9. subject unique identifier (versions 2 and 3 only)
10. extensions (version 3 only)
11. signature on the above fields

Let us consider two users U_1 and U_2 subscribed to a CA. Before communicating with each other, U_1 requests U_2's certificate from the CA. When U_1 gets the certificate of U_2, it comes to know about the public key of U_2 by decrypting it with its own public key as the certificate is signed by secret key of CA. Henceforth, if a message from U_2 to U_1 is signed by U_2's private key, then U_1 can understand the authenticity of the message and the sender (U_2).

If U_1 and U_2 subscribed to two different certification authorities CA_1 and CA_2 respectively, then before communicating with each other, U_1 and U_2 attempt to verify the authenticity of the other CA. For example, CA_2 secretly shares its certificate with CA_1, and then CA_1 signs CA_2's certificate and shares with U_1. Then U_1 can get hold of CA_2's public key and hence can authenticate any user like U_2 who has subscribed to CA_2.

X.509 also defines syntax for certificate revocation lists (CRLs). A certificate revocation list is a list of all certificates freshly issued by a CA before their older certificates expired. Thus a CA, upon requests for new certificates, always checks its CRL first to make sure the latest version is either not present or has already expired.

Authentication in X.509 can be achieved using one-way, two-way or three-way handshake mechanisms. The updates included in v3 accommodate provisions for sharing extended information like network ID, message subject and user-related information.

4.3 ELECTRONIC MAIL SECURITY

4.3.1 Overview

Electronic mail is the most popular platform for exchanging business information across the internet or any internal network within a company. It is supported by almost all operating systems and hardware frameworks. At its basis, the electronic mail requires two principal components—mail servers, hosts that deliver, forward, and store emails; and mail clients, software applications that interface with users and allow them to read, compose, send, and receive emails.

To support their basic functional need, mail servers routinely connect to unreliable network sources. They receive mails from unknown users and deliver them to other unknown users. E-mails are thus most prone to intentional or unintentional leakage of confidential information. As the mail server cannot and does not assume responsibility for unknown email addresses, outsiders can use them to spread virus, Trojan horses or malicious code in the internal network. Mail servers are mostly public, so users can easily attack them and compromise their security, thereby indirectly getting access to internal networks.

As mail servers often need to penetrate the internal network's barrier of defence, it becomes a very handy tool for attackers who can launch attacks against the internal network using such penetrations. Being one of the most heavily used network applications, a mail server's internal architecture and design are well-known to almost all network users. It is thus easy for an attacker to leverage such information, especially the design flaws of the mail server, to launch an attack. Therefore, mail server administrators need to adequately understand and deal with the security threats inherently involved and also deploy measures to protect the internal network from such threats.

4.3.2 Pretty Good Privacy as a Solution to E-mail Security

4.3.2.1 Overview

Pretty Good Privacy (PGP) is an encryption-decryption-based algorithmic approach to provide privacy and authentication in data communication. PGP is often used for signing, encrypting and decrypting e-mails to enhance the security of e-mail communications.

PGP was invented by Phil Zimmermann, who wanted to provide a platform-independent solution for email communication that can be implemented as a set of simple commands. PGP is well-documented for easy understanding and deployment, while remaining flexible enough to accommodate any cryptographic algorithm as its basis. These simple yet powerful design principles helped make PGP an extremely popular choice among several different organizations.

4.3.2.2 What exactly does PGP do?

PGP encryption is a well-defined set of steps. The steps include hashing, data compression, symmetric-key cryptography and public-key cryptography. Each step is flexible enough to accommodate any one of the several supported algorithms.

PGP can provide five services—authentication, confidentiality, compression, e-mail compatibility, and segmentation, as follows:

- ***Authentication***

PGP can be used to authenticate a message sent by a user using hashing and public key cryptography:

1. The sender creates a message and generates a digital signature for the message.
2. A hashing function (e.g., SHA-1) is used to generate a 160-bit hash of the message.

3. The generated hash is signed with the sender's private key, and attached to the message.
4. Upon receiving the message with the signed hash, receiver decrypts the hash with the sender's public key.
5. Receiver verifies the received message by re-hashing it and comparing the new hash with the decrypted hash.

- ***Confidentiality***

PGP provides confidentiality with private key cryptography by generating a one-time key. This 128-bit (16 bytes) key is created randomly and used only once as a secret key to encrypt a message. The basic mechanism is as follows:

1. The sender creates a message and generates a128-bit random number as the session key.
2. The sender encrypts the message using CAST-128 / IDEA / 3DES in CBC mode with the session key.
3. The sender encrypts the session key with the recipient's public key and attaches the former to the encrypted message.
4. The receiver decrypts the session key with its own private key.
5. The receiver decrypts the message with session key.

- ***Compression***

Message compression is yet another service provided by PGP. Compression usually happens after generation of signature from the message and before encrypting it. PGP generates the signature from the uncompressed message because it is more convenient to store a signature with uncompressed message than with a compressed one and it is also much simpler to validate an uncompressed message with the stored signature than it is to validate a compressed message. Whereas compression is carried out before encryption to complicate the process so that encryption procedure strengthens itself.

- ***Email Compatibility***

PGP deals with binary data, whereas most mail clients only permit ASCII text. To provide compatibility between these two formats, PGP features a special service for encoding raw 8-bit octets into a stream of printable ASCII characters (using an algorithm like Radix-64). This feature also appends a Cyclic Redundancy Check (CRC) at the end of an octet to detect transmission errors. See Stallings Appendix 15B for a detailed description of this feature.

- ***Segmentation***

If the message to be transmitted is too long for a single email, PGP segments it into smaller chunks. Segmentation is done after all other steps have been performed (even after the compression and application of Radix-64). Upon receiving these smaller chunks, PGP at the receiver side removes mail headers and recombines the segments. Then it applies all other steps in the reverse order and gets back the original message.

4.4 IP SECURITY

4.4.1 Overview

Although we saw a plethora of services that provide security for the application layer and the transport layer, we did not see much in terms of security for the lower layers. One reason behind such general lack of lower-layer security mechanisms is that the Internet Protocol on its own does not provide any comprehensive security framework. But there have been attempts to implement security controls in the IP layer, so that an organization cannot only ensure security for applications that already have security mechanisms in place, but it can also ensure security for those applications which do not have inherent security control.

Since IP datagram travel between unknown nodes in a potentially unsecured network, any information contained in them is liable to eavesdropping and even alteration. Thus, security control in the IP layer has to ensure confidentiality of IP datagram in transit as well as the authenticity of the origin and that of the datagram. IP Security (IPSec) is a protocol suite established as an extension of the Internet Protocol. It defines a framework for addressing the security issues in the IP layer and accommodates new security solutions.

IPSec is a successor of the Network Layer Security Protocol (NLSP). It ensures confidentiality by encrypting each IP packet. It can be used between two hosts or two network gateways or between a network gateway and a host. IPSec also includes provisions for establishing mutual authenticity between end-points and sharing of secret keys for a particular session between end-points.

4.4.2 Understanding the IPSec Architecture

The IPSec architecture is fairly complex as the protocol suite provides for implementation of confidentiality as well as authentication. The IPSec protocol suite includes five principal components—Authentication Header (AH), Encapsulating Security Payload (ESP), Internet Key Exchange (IKE), Internet Security Association and Key Management Protocol (ISAKMP)/ Oakley, and transforms. In order to understand, implement, and use IPSec, it is necessary to understand the relationship among these components, how they interact with each other and how they are tied together to provide the capabilities of the IPSec suite.

The IPSec architecture defines the capabilities of hosts and gateways, individual protocol, datagram header formats for the protocols. The architecture also discusses the semantics of the

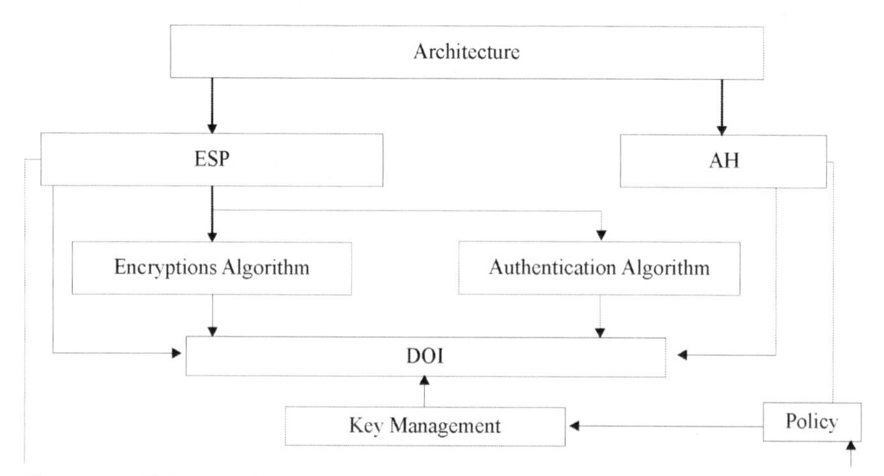

Figure 4.2: IPSec Roadmap.

IPSec protocols and the issues involved in the interaction between several IPSec protocols and the rest of the TCP/IP protocol suite.

- ***Architecture***

Covers general concepts, definitions, security requirements and mechanisms; no specific details of individual segments are given.

- ***Encapsulating Security Payload (ESP)***

It provides confidentiality for datagram by encryption, encapsulation and optionally, authentication. ESP defines the format of a generic IPSec packet and discusses the issues involved.

- ***Authentication Header (AH)***

It describes packet structure, payload header format and mechanisms for data content and origin authentication as well as connectionless data integrity for IP packets.

- ***Encryption Algorithm***

These are a set of specific transformations used by ESP for enforcing confidentiality. It includes documents that talk about algorithms involved, key sizes, key generation and the overall transformation mechanism.

- ***Authentication Algorithm***

It includes documents that describe how various authentication algorithms are used by AH and the authentication option of ESP.

- ***Domain of Interpretation (DOI)***

It defines payload formats, exchange types, and several other operational parameters.

- ***Key Management***

It is a set of documents describing several key management schemes.

4.4.3 IPSec Implementation

Depending on the security requirements of users, IPSec can be implemented and deployed in the end hosts or in the gateways/routers or in both. These two types of deployment have their own merits and demerits as they address two different security goals. The host implementation is most useful when end-to-end security is desired. The router implementation is useful when security is desired over a part of a network.

- *Host Implementation*

A "host" in this context refers to the device where a packet has originated. The host implementation of IPSec has the following advantages:

- o Provides end-to-end security
- o Covers all modes of IPSec
- o Provides security on a per flow basis
- o Maintains user context for authentication (in establishing IPSec connections)

Host implementations can be classified as follows:

o OS-Integrated

This type of host implementation is integrated with the operating system. As IPSec protocol suite is an extension of IP protocol, the former can be designed as part of the network layer (i.e., IP layer).

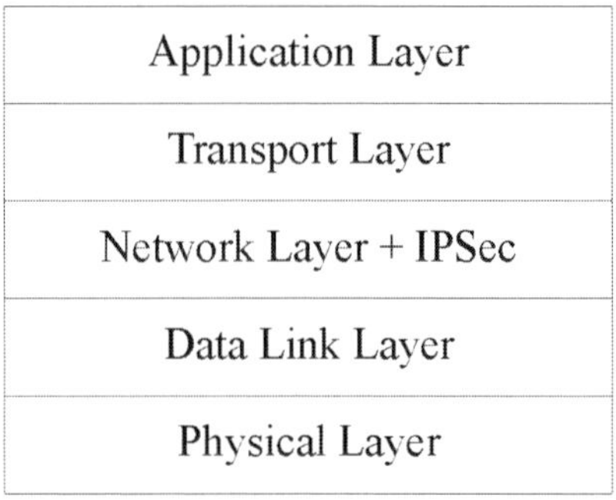

Application Layer
Transport Layer
Network Layer + IPSec
Data Link Layer
Physical Layer

Figure 4.3: IPSec stack layering.

This implementation is straight-forward and resembles ICMP. It comes with a set of advantages:

- Because of the tight integration with network layer, IPSec can access various network layer services.
- It is easier to provide per flow security, because key management, the base of IPSec protocols, can be seamlessly integrated with the network layer.
- All IPSec modes are supported.

o Bump in the Stack (BITS)

In this host implementation, a separate IPSec layer is introduced between the network layer and the data link layer. For companies providing VPN and intranet solutions, the OS-integrated implementation had a serious drawback in the sense that the companies were required to comply with features provided by the OS vendors, something that had limited the former's capabilities to provide advanced solutions. The BITS implementation addressed this issue.

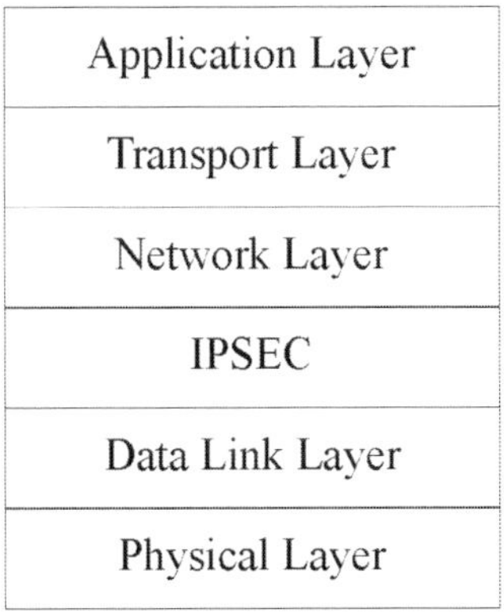

Application Layer
Transport Layer
Network Layer
IPSEC
Data Link Layer
Physical Layer

Figure 4.4: BITS IPSec stack layering.

Unlike the OS-integrated implementation, BITS specifies most of the network layer features (e.g., route tables, fragmentations, etc.), which has led to duplication of effort and raised several unwanted and unmanageable complications. This is a major drawback of BITS. But otherwise, BITS can give a better host implementation by allowing integration with specialized firewalls, especially in the case of intranets and VPNs.

- ***Router Implementation***

The router implementation of IPSec is meant to provide the ability to secure a packet over a part of a network, which is achieved by *tunnelling* the packets through that part.

The router implementation has the following advantages:

- o Ability to secure packets flowing between two hosts over a public network such as the internet.
- o Ability to authenticate and authorize users entering a private network, which enables many organizations to allow their employees to telecommute over the internet via the former's VPN or intranet.

Router implementations can be classified as follows:

o Native Implementation

This is analogous to the OS-integrated host implementation; the only difference is that now the integration is with routers instead of operating systems.

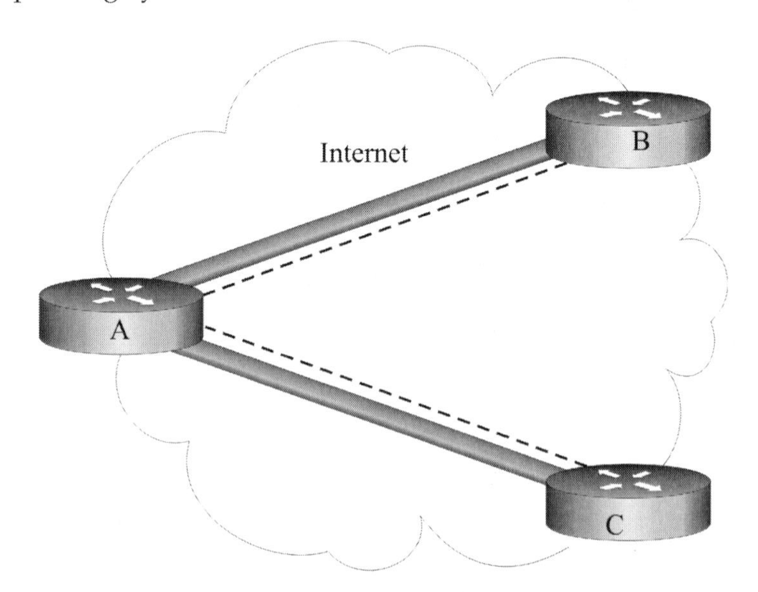

Figure 4.5: A Native Implementation Architecture.

o Bump in the Wire (BITW)

This is analogous to the BITS host implementation. In this variant, IPSec is incorporated in a hardware device that is attached to the physical interface of a router. This router normally does not run any routing algorithm and is used for the sole purpose of securing packets. BITW is not a long-term solution, because it is not economically feasible to have a hardware device attached to every inerface of every router.

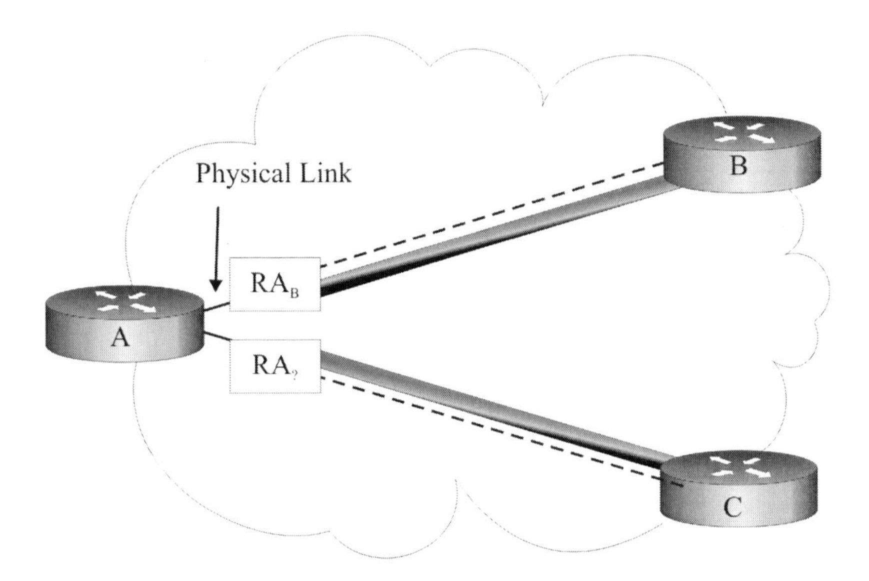

Figure 4.6: BITW Deployment Architecture.

4.4.4 Security Association

Security Association (SA) is an important concept in IPSec for both confidentiality and authentication. SA can be viewed as a secured unidirectional contract between two communicating parties. It is used between two communicating peers to authenticate an IP datagram sent from the sender and also to ensure confidentiality of the contained data. SA operates in the simplex mode of IPSec, i.e., it is one-way. Hence if in a situation peers need a two-way secured exchange, then two SAs will have to be set up.

SA is also protocol-specific, i.e., there must be an SA defined specifically for each individual protocol in IPSec. For example, if two peers support authentication using both AH and ESP, then there should be two separate SAs to support each protocol. All IPSec implementations build and maintain an SA Database (SADB) that stores the IPSec protocols and their respective SAs.

The Security Policy Database (SPD) is another important component of IPSec architecture which works in tandem with the SADB. It defines the security policies for each protocol and their corresponding IP packets.

An SA is uniquely identified by its three major components:

- Security Parameters Index (SPI),
- IP Destination address and
- Security Protocol Identifier.

Being a unidirectional contract between two peers, it is important for a source to identify an SA. This is achieved by the SA Selector. For the destination node, determining the authenticity of the sender node and confidentiality of the data received in an SA contract can be a bit tricky. This is addressed by sending the unique identifier of SA with each packet. The 32-bit unique identifier is known as the Security Parameters Index (SPI). The SPI is passed as part of AH and ESP headers.

4.4.5 Authentication Header

Authentication Header defines the provisions for source authentication of IP datagram and ensuring data integrity in IPSec architecture. The authentication mechanism involves the use of message authentication code (MAC), which in turn necessitates the use of secret key sharing between communicating end-points. This feature protects the network from an IP spoofing attack, while allowing workstations to authenticate the sender and filter packets accordingly. The data integrity provision of AH ensures zero possibility of undetected alteration of datagram content in transit.

The authentication header consists of the following fields:

- **Next Header:** 8 bits that identify the IP header type.
- **Payload length:** 8 bits that contain the total length of authentication header minus 2.
- **Reserved bits:** 16 bits, reserved for future extensions.
- **SPI:** 32-bit unique identifier for Security Association.
- **Sequence number:** A monotonically increasing unique identifier (32 bits).
- **Authentication data:** A variable-length field containing several parameters (e.g., MAC, Itegrity Check value (ICV), etc.).

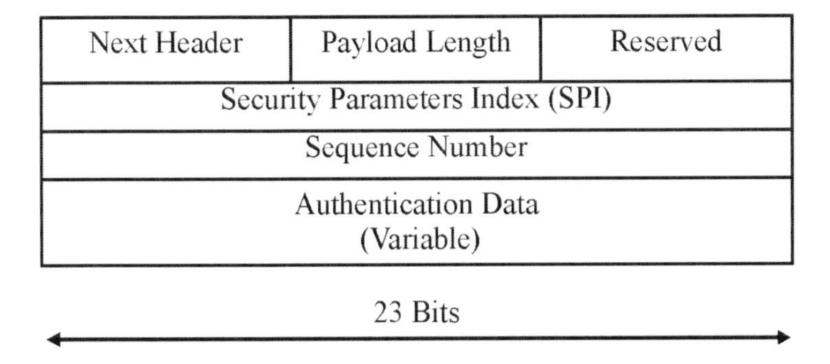

Figure 4.7: Authentication Header Packet Format.

Authentication using AH can be realized by computing an authentication function on the IP datagram using a secret authentication key. Certain fields, which are bound to change in transmission (e.g., Hop count), are excluded from this computation. Although there is a wide array of authentication algorithms available in the market, IPSec architecture mandates the use of two mechanisms—HMAC-MD5 and HMAC-SHA1.

After authenticating an IP datagram, data integrity can be ensured using the Integrity Check value (ICV). ICV is a checksum over the message content (by the authentication function described earlier). The Sequence Number, a monotonically increasing unique identifier, is used to rearrange IP packets at the receiving end. It gives protection against duplication of datagram, thereby providing security against replay attacks.

4.4.6 Encapsulating Security Payload (ESP)

Although an extended version of ESP provides authentication services like AH, the major role played by ESP in the IPSec architecture is to provide confidentiality services including secrecy of datagram content and limited traffic confidentiality.

The ESP packet format can be viewed as a composition of the following parameters:

- **SPI:** 32-bit unique identifier for an SA.
- **Sequence Number:** (Same as in AH).
- **Payload data:** A variable-length field containing encrypted IP datagram or transport layer segment.
- **Padding:** May range from 0 to 255 bytes. It hides the actual length of payload data and also helps pad data blocks to support encryption into fixed-size cipher blocks.
- **Pad length:** These 8 bits give the length of the previous padding block.
- **Next Header and Authentication Data:** (Same as in AH).

Confidentiality can be achieved in ESP by encrypting the payload, Padding, Pad length and/or the Next Header using any of the supported standard encryption mechanisms like 3DES, RC5, IDEA, 3IDEA, etc.

4.4.7 IPSec Operation Modes

IPSec can operate in two modes—transport mode and tunnel mode. In IPSec architecture, protocols like AH and ESP implements both the tunnel and transport modes. Thus there are four possible combinations of modes and protocols. In practice, AH in tunnel mode is never used because it gives the same protection as AH in transport mode.

The AH and ESP headers do not change between tunnel and transport modes. The difference between these two modes is in what it is that they are protecting, the IP packet or an IP payload.

- ***Transport Mode***

The transport mode is used only when the security need is end-to-end. In practice, routers look into the network layer to make routing decisions. They can at most change network layer headers and nothing beyond that. The transport mode violates this general rule by forcing routers to append an IPSec header to a datagram on-the-fly.

In the more usual case, when end-to-end security is not desired, TCP/UDP packets flow directly into network layer from the transport layer. But in transport mode, the transport header intercepts the packet flow between transport and network layers.

Let us consider the following example of two hosts, A and B, which are configured for IPSec Transport mode to provide end-to-end security. When data security is the only concern, transport mode of ESP is sufficient. But if authentication is needed as well, then AH should be deployed (although AH andESP can be used together in this case).

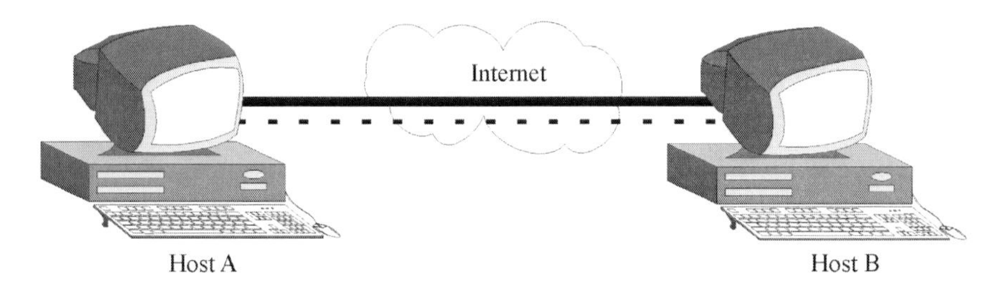

Figure 4.8: Hosts with Transport ESP.

For the OS-integrated host implementation, packets generated in transport layer flows into the IPSec component where AH and ESP headers are attached to them. Then they flow into network layer for appending of the IP header.

If both AH and ESP headers are appended, then we need to be careful regarding the order of the appending. If ESP is appended after AH (Fig. 4.9), then authentication and integrity are limited to only the transport layer payload, which is not usually desired.

IP	ESP	AH	TCP hdr	Data

Figure 4.9: Packet format with ESP and AH.

If AH is appended after ESP (Fig.4.10), then both security as well as integrity of a larger chunk is ensured—thanks to ESP for larger chunksand thanks to AH for integrity insurance.

IP header	AH header	ESP header	TCP payload

Figure 4.10: Packet format with AH and ESP.

The BITS implementation of transport mode is not as efficient and user-friendly as the OS-integrated implementation. In BITS, the ESP and AH headers are appended after the IP payload is formed, which means BITS will have to re-implement the IP functionality for calculating the IP checksum and fragmenting the packet if necessary. This is why many BITS implementations do not support transport mode and limit themselves to tunnel mode.

- ***Tunnel Mode***

The tunnel mode of IPSec is generally implemented within routers, much like BITW. It is effective only when end-to-end security is not desired, i.e., the security end point is different from the ultimate destination (or when the destination is not known in advance). It is widely used by VPNs and by devices that are meant for security, not packet handling.

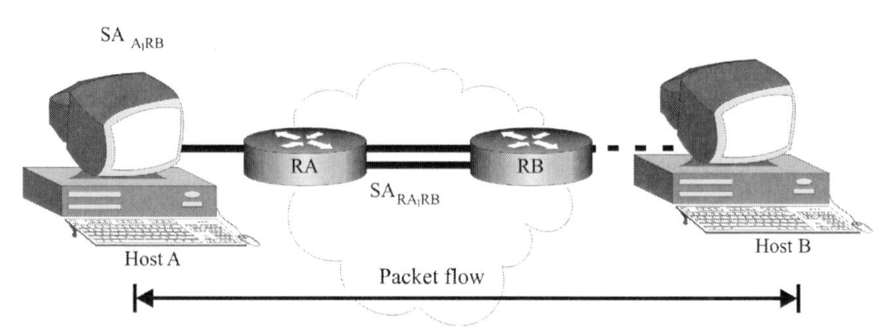

Figure 4.11: IPSec in tunnel mode.

In contrast with the transport mode, two IP headers are appended in the tunnel mode—an inner header and an outer header. The inner one is appended by the IP Layer. Then the device (which provides security) appends an IPSec header and the outer IP header to the IP payload. Hence, it is easy to understand that the transport mode is more efficient than the tunnel mode since the latter involves more processing.

IP header	ESP	IP header	Network payload

Figure 4.12: IPSec tunnelled mode packet format.

IPSec defines tunnel modes for both AH and ESP. IPSec also supports nested tunnels where a tunnelled packet is tunnelled again (Fig. 4.13).

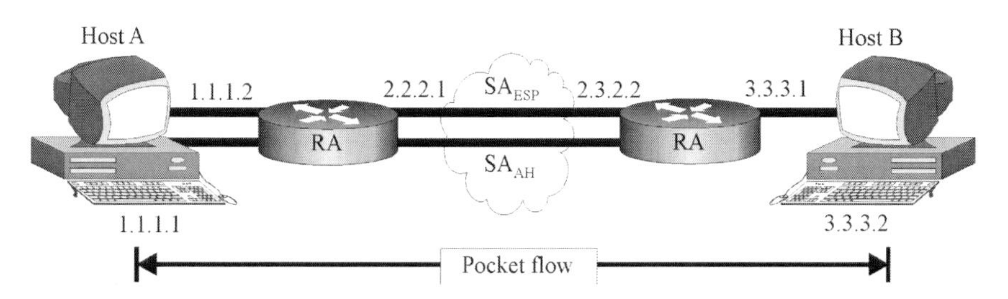

Figure 4.13: A Nested tunnel example.

IP header	ESP	IP header	AH	IP header	Data
Src = 2.2.2.1 Dst = 2.3.2.2		Src = 1.1.1.1 Dst = 2.3.2.2		Src = 1.1.1.1 Dst = 3.3.3.2	

Figure 4.14: B Nested packet format.

For example, let us consider host A is sending a packet to host B (Fig. 4.13) A has to authenticate itself to router RB and there is a VPN between the two networks bordered by RA and RB. When the packet is seen by router RB, its outermost header describes it as a tunnelled ESP packet. The tunnelled packet is carrying a tunnelled AH packet, which in turn is carrying the IP packet destined for host B generated by host A.

IP header	ESP	IP header	Network payload

Figure 4.15: Valid tunnel.

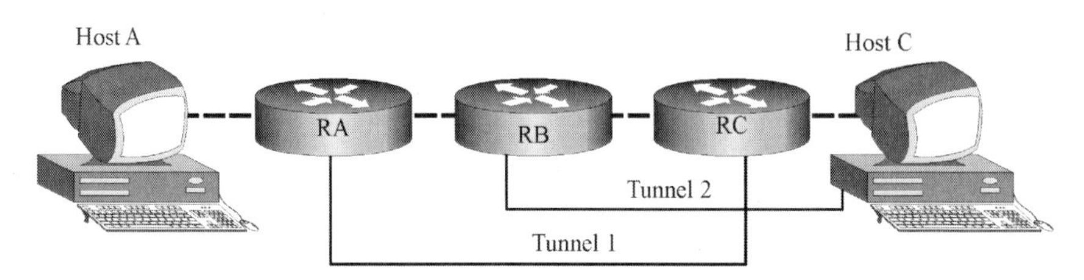

Figure 4.16: Invalid tunnel.

The requirement for a valid tunnel (Fig. 4.15) is that the inner IP header be completely encompassed by the outer IP header. Two tunnels must not overlap each other. If they do, then they are not valid nested tunnels (Fig. 4.16).

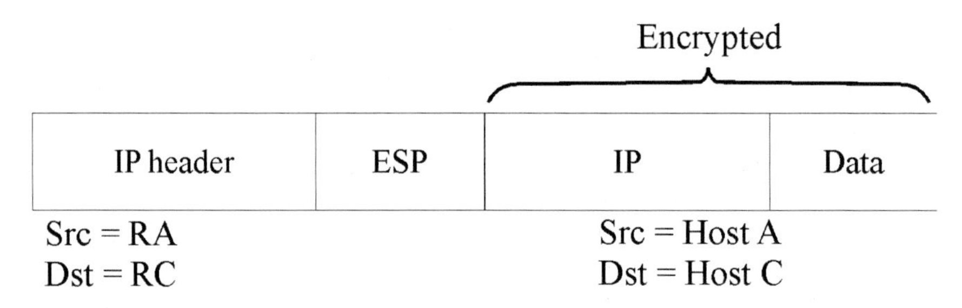

Figure 4.17: Valid Tunnelled Packet.

To better understand valid nested tunnels, let us look into the flow of a packet. First, RA constructs a tunnelled packet as shown in Fig. 4.17. When the packet reaches RB, if the latter tunnels the packet to host C, then after leaving RB, the packet format is as shown in Fig. 4.18.

IP header	ESP	IP header	ESP	IP	Data
Src = RB Dst = Host C		Src = RA Dst = RC		Encrypted	Encrypted

Figure 4.18: Invalid Tunnelled Packet.

It is straight-forward to notice that this format is incorrect because the packet now reaches host C before reaching RC. When the packet reaches host C, the latter processes the AH header. When the second IP header is exposed, host C drops the packet because the destination is specified as RC and not itself. Hence it is clearly understood that nested tunnels are difficult to build and maintain and should be used sparingly.

4.4.8 Key Management

Key management in IPSec architecture refers to the mechanism for computing keys and distributing them among communicating agents. Like both AH and ESP, SA establishment also involves key generation and distribution, which can be addressed by the key management part of IPSec.

This feature talks about two major modes of key exchange:

- **Manual:** This is old-fashioned, but is widely in use. Here, network administrators manually configure keys of all communicating parties. This is only feasible for small networks containing at most a few hundred systems.
- **Automatic:** This is meant for on-demand and dynamic generation of keys and their distribution across a large network involving thousands or millions of communicating peers. The default automatic key exchange protocol for IPSec is the Internet Key Exchange (IKE).

Internet Security Association and Key Management Protocol (ISAKMP) define a framework for internet key management. It does not specify the use of a particular algorithm for key exchange, and remains generic enough by only discussing security protocols, message formats, attributes and parameters necessary to negotiate security keys across unreliable networks. Initial versions of ISAKMP use the Oakley Key Determination protocol, which is itself a refinement of the Diffie-Hellman Key Exchange. In later versions, IKE is used as the standard key exchange mechanism. IKE uses a hybrid of Oakley and SKEME protocols.

4.5 WEB SECURITY

4.5.1 Overview

In today's world it is impossible to think of the existence of an organization that has no web presence. Ranging from the government to small-scale businesses and individual users, everyone seems to possess their own website. Not only is it a very popular means for communicating with customers or sharing opinions and individual thoughts, but it is also becoming increasingly more common to do the daily shopping and transactions on the web. This has revolutionized the way businesses are operated. With more demanding customers and more competition from all sides, companies were obliged to come up with advanced features like augmented reality, graphical web applications and online shopping carts. These complex applications led to a complete revolution about the way security was traditionally perceived and managed. This unimaginable complexity and extreme profit-making potential make the web one of the most vulnerable entities around.

An unsuspecting user, who is completely unaware of the technology being used by the website (s)he is browsing, may be suddenly exposed to a great vulnerability and his/her secret information may suddenly be compromised by a third party. An attacker may intentionally hide his/her malicious code in a website and may trick users to trip over that code again and again, every time losing yet another piece of sensitive information to the attacker. The possibilities are endless and enormous. An attacker can gain knowledge about information like:

- What operating system, technology framework and application platforms are in use by the user workstation.
- What type and what version of web browser are in use and what are the add-ons installed.
- Browsing history and surfing trend of a user.
- User credentials stored in local memory in the form of *cookies*.

- Web application and network configurations like IP addresses, host addresses, protocols in use, etc.

Thus web security is challenging and extremely important. No wonder it has been an area of intense research. Web security has long become the security focus for application administrators and network managers.

4.5.2 Web Security Threats

The main threats to web systems are:

- **Physical:** Physical threats like fire, water, smoke, dust, theft and any other physical damage to the network devices and equipments constitute a set of very crude means for causing damage. These threats can be brought about by an unintentional or a malicious physical intervention, and can result in server break-downs, low system uptime and many other devastating consequences. If the power management and backup plans are not so good, then power break-downs add another source of concern to the problems cited above.
- **Human error:** Human errors like manhandling of systems by operators, flaw in data setup by database administrators, mal-configuration of web servers and security controls, even improper handling and configuration of hardware devices can be disastrous for system performance and may expose the system to unforeseen vulnerabilities. Be it unintentional or pre-meditated, human errors can snowball to cause authentication errors, poor data validation, improper confidentiality checks and intrusion threats. Planned network management, configuration control, good programming practices and user awareness and training programs are some of the steps that reduce the possibility of human errors.
- **Malfunction:** Both hardware and software malfunctions can adversely impact the operations of a website or web application, thereby causing huge financial losses. One way to (partially) avoid such malfunctions is to introduce *redundancy* into the system. For example, instead of one

dedicated server for handling user requests, multiple servers working in parallel (or multiple data centers) should be employed.

- **Malware:** Malwares are malicious software launched by internal or external attackers. They come in the form of Trojan horses, virus, worms, etc. They can be injected into the system via web applications or search engine optimizing software (e.g., spambots).
- **Spoofing:** Spoofing is a technique where the attacker obtains knowledge of valid access credentials (e.g., username of a user or IP address of a machine) and then pretends to be the owner of such credentials (e.g., a valid user or a valid workstation). Under such a guise, it is only a matter of time before a system gets compromised by the attacker. Once the attacker gets hold of system resources and classified data, (s) he can even launch many other related attacks like scanning, scavenging, eavesdropping and DoS.
- **Scanning:** It is a methodology to learn about a network and user information by monitoring web resources. It is usually achieved by fingerprinting web applications or network controls prior to an attack, but can also include brute force or dictionary mechanisms to glean secret user information.
- **Eavesdropping:** Eavesdropping refers to secret monitoring of shared information. The information can either be shared through a network or it can be shared via exchanges in the request-response cycle of serving a webpage. Eavesdropping can be used to obtain a lot of private information. The attacker can cause the information to leak out to the public, thereby causing devastating consequences.
- **Scavenging:** In a scavenging attack, the attacker carefully analyzes the data (s)he successfully eavesdropped and thus tries to deduce further information regarding the underlying business logic or operational logic.
- **Spamming:** Spamming refers to overloading a web server with junk data, which usually leads to DoS conditions.
- **Out of band:** Tunnelling can be maliciously leveraged to access the lower levels of protocol stack. This is known as

an "out of band" attack. It can be used to gain control over routers and their functional modules.

4.5.3 Overview of Security Threat Modelling and General Counter-measures

Web application security must be addressed across different levels of the protocol stack. A small loophole in any one level may render the whole website surprisingly vulnerable. The figure below (Fig. 4.19) shows the scope of web security in a three-layered architecture. It also depicts threat modelling and security practices in each layer. Threat modelling refers to securing the network, securing the host, and securing the application.

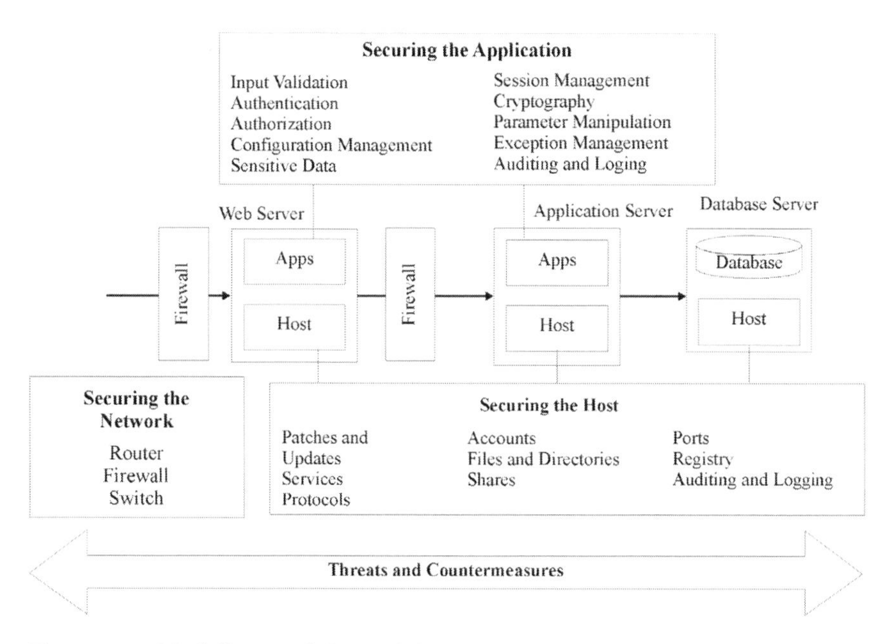

Figure 4.19: Modelling and General Counter Measures.

A major challenge associated with web security is to ensure the confidentiality of private information. It can be addressed with the use of *Anonymizers*. There are many other applications that address generic web security threats. The Secure Socket Layer (SSL) and the Secure Electronic Transaction (SET) are two of them.

4.5.4 Secure Socket Layer and Transport Layer Security

4.5.4.1 Overview

The SSL protocol was originally introduced by Netscape to ensure security of data transported and routed through HTTP, LDAP and POP3 applications. The latest version of SSL (v3) is a result of standardization by the Transport Layer Security (TLS) group of Internet Engineering Task Force (IETF). They took public feedback and industry reviews into account while coming up with SSL v3. SSL/TLS is designed to establish an end-to-end secured channel in a client-server environment. SSL works at the TCP layer. Although it was originally meant to provide confidential exchange of data in any application, it has been mostly used for HTTP applications. A majority of present-day web servers provide support for SSL/TLS. Netscape and Microsoft IE even came up with their own SSL-enabled client applications.

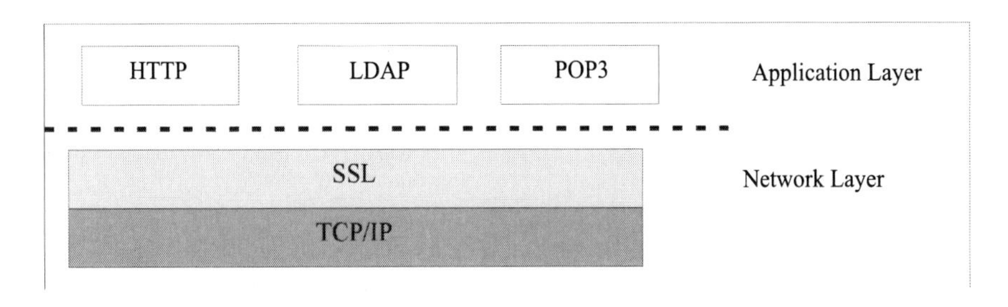

Figure 4.20: SSL between application protocols and TCP/IP.

4.5.4.2 SSL Design Goals

SSL design goals are as follows:

- Should support both client and server with the provision to authenticate each other. Although user authentication is mostly done by certification mechanisms like X.509, SSL provides a baseline option by using the most general cryptographic techniques.

- Should support confidentiality of data. It is one of the major design objectives of SSL. SSL not only ensures secrecy of user data in transit, but it also ensures confidentiality in the secret key negotiation process.
- Should ensure that data in transit can neither be snooped nor altered by any means.

4.5.4.3 Architectural details and implementation

The SSL protocol suite can be broadly classified into two sections:

- Protocols like SSL Handshake Protocol, SSL ChangeCipher Spec protocol and SSL Alert Protocol. They enable the establishment of an SSL Secured Channel and perform several functional requirements like session management, data transfer between peers, cryptographic parameters handling, etc.
- SSL Record Protocol. It is meant to ensure data security and integrity. It is also used to encode data sent by other protocols in the suite.

The SSL protocol stack can be explained with the following diagram (Fig. 4.21).

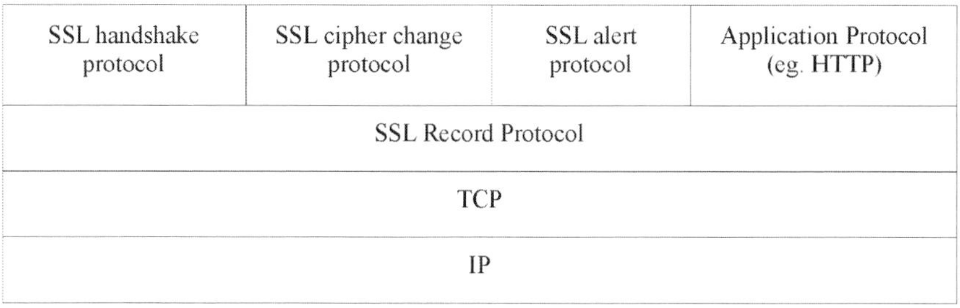

SSL handshake protocol	SSL cipher change protocol	SSL alert protocol	Application Protocol (eg. HTTP)
SSL Record Protocol			
TCP			
IP			

Figure 4.21: The SSL protocol stack.

- ***SSL Session and Connection***

Two major concepts associated with SSL are Sessions and Connections:

- **Connection:** A connection is a logical link between client and server with the provision of a particular type of service. In SSL terms, it must be a peer-to-peer connection with two network nodes. Each connection must be associated with a session.
- **Session:** It defines an interaction between a client and a server identified as a set of parameters such as algorithms used, session number, etc. An SSL session is established using the Handshake Protocol. The handshake protocol allows parameters to be shared using the connection made between the server and the client. Sessions are also used to avoid negotiation of new parameters for each connection. So, a single session can be shared among multiple SSL connections between the client and the server. In theory, it may also be possible that multiple sessions are shared by a single connection, but this feature is not used in practice.

SSL sessions and connections are characterized by several parameters. Some of them are used during the negotiations of the handshake protocol, while some others are used in establishing encryption methods. A session is defined by the following parameters:

- Session identifier: This is an identifier generated by the server to identify a session with a chosen client,
- Peer certificate: X.509 certificate of the peer,
- Compression Method: A method used to compress data prior to its encryption,
- Cipher Spec: It specifies the data encryption algorithm (e.g., DES) and the hash algorithm (e.g., MD5) used during the session,
- Master Secret: 48-byte data, secretly shared between the client and the server,
- IsResumable: This is a flag indicating whether the session can be used to initiate new connections.

According to specification, an SSL connection state is defined by the following parameters:

- Server and client random: Random data generated by both the client and the server for each connection,

- o Server write MAC secret: The secret key used for data written by the server,
- o Client write MAC secret: The secret key used for data written by the client,
- o Server write key: The bulk cipher key for data encrypted by the server and decrypted by the client,
- o Client write key: The bulk cipher key for data encrypted by the client and decrypted by the server,
- o Sequence number: Sequence numbers (maintained separately by each party) keep track of individual messages transmitted and received during a particular session.

• ***The SSL Record Protocol***

The SSL Record protocol provides Confidentiality and Message Authenticity—two specialized services of the SSL Protocol Suite. SSL Record protocol is used by the upper level protocols of SSL suite, e.g., the handshake protocol. The upper level protocols send messages to the Record Protocol for processing, which involves a number of steps.

The message is first segmented into manageable portions that can be handled within a single delivery. Each segment is known as a Record Protocol Unit.

These segments undergo an optional lossless compression algorithm, (SSL v3 and latest SSL/TLS do not enforce such compression) followed by an optional padding.

Then a Message authentication Code (MAC) is computed on the message which is later used to verify the integrity of the message.MAC is computed as a hash function (e.g., MD5 or SHA-1):

MAC = Hash function [secret key, primary data, padding, sequence number].

and then attached to the primary data. The secret key used above is either a client write MAC secret or a server write MAC secret, depending on which party prepared the packet. The length of MAC value depends on the hash function used. After receiving the packet, the receiving party computes its own MAC value and compares it with the received MAC value. If two MAC values match, the data has not been modified in transit.

Next, the data plus the MAC value are encrypted together using a standard symmetric encryption algorithm (usually DES or triple DES). Finally, a header consisting of the following fields is attached to the encrypted data:

- o Content type: 8 bits that identify the type of payload in the packet. Content type determines which higher-level protocol to use for processing the packet. The possible values are change_cipher_spec, alert, handshake and application_data.
- o Major version: 8 bits that specify the major protocol version. For example, for SSL 3.0, the value is 3.
- o Minor version: 8 bits that specify the minor protocol version. For example, for SSL 3.0, the value is 0.
- o Compressed Length: 16 bits that give the length of the message—either the original message or the message after compression if compression was involved.

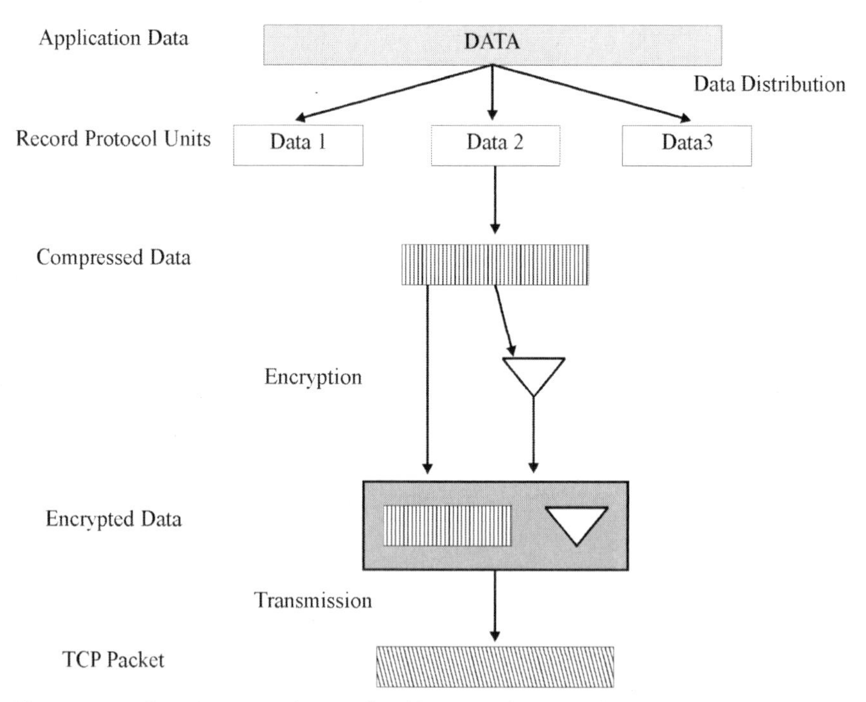

Figure 4.22: Creating a packet under SSL record protocol.

After appending of the header, primary data plus MAC constitute an object known as an SSL Record. This is ready for transmission using the TCP protocol.

- ***The Change Cipher Spec Protocol***

This protocol is the simplest of the three SSL protocols that use SSL Record Protocol. The Change Cipher Spec Protocol consists of a single message that carries the value 1, which is transferred using the SSL Record Protocol. The sole purpose of this message is to cause a pending session state be established as a fixed state, which results in an updating of the cipher suite to be used. This type of message must be sent by a client to a server and vice versa, before a session state is considered as agreed.

- ***The Alert Protocol***

As the name suggests, the Alert Protocol is used to share alert messages among peers. It is used by different parties to convey SSL session messages associated with data exchange and functioning of the protocol. Each message in the alert protocol is 16 bits (two bytes) in size. The first byte is the alert value (e.g., "warning (1)" or "fatal (2)") that determines the severity of the message sent. For example, sending a message having a "fatal" status by either party will result in an immediate termination of the corresponding SSL session. The second byte of the message contains one of the defined error codes, which may occur during an SSL communication session.

- ***The Handshake Protocol***

The Handshake Protocol is the most complex part of the SSL Protocol Suite. It is used to initiate a session between a server and a client, and to negotiate, among others, the algorithms and keys used for data encryption. This protocol makes it possible to authenticate the parties to each other and negotiate appropriate session parameters. The process of negotiation between the client and the server can be divided into 4 phases:

o Establish Security Capabilities

During the first phase, a logical connection is initiated between the client and the server followed by negotiation on the connection parameters. The client sends the server a "client_hello" message containing data such as:

§ Version: The highest SSL version supported by the client.

§ Random: Data consisting of a 32-bit timestamp and 28 bytes of randomly generated data. This data is used to protect the key exchange session between the parties.

§ Session ID: A number that defines the session. A nonzero value in this field indicates that the client wants to update the parameters of an existing connection or establish a new connection on this session. A zero value in this field indicates that the client wants to establish a new connection.

§ Cipher Suite: A list of encryption algorithms and key exchange methods supported by the client.

The server, in response to the "client_hello" message, sends back a "server_hello" message containing the same set of fields:

§ Version: The lowest version number of the SSL protocol supported by the server.

§ Random Data: Same as that from the client, but the random data is completely independent (and of course the timestamp is different as well).

§ Session ID: If the client field was nonzero, the same value is sent back; otherwise, this field contains the ID for a new session.

§ Cipher Suite: The server uses this field to send back a single set of protocols that was selected by the server from the ones proposed by the client. The first element of this field is a chosen method of key exchange between the client and the server. The next element specifies the encryption algorithms and hash functions that will be used during the session, along with all of their parameters.

The set of encryption algorithms and key exchange methods sent in the CipherSuite field has three components:

1. The method of key exchange between the server and the client,
2. The encryption algorithm for data encryption,
3. A hash functions for obtaining the MAC value.

o Server Authentication and Key Exchange

The server begins the next phase of negotiation by sending its certificate to the client for authentication. The message sent to the client contains either one X.509 certificate or a series of X.509 certificates. The certificates are necessary for authenticating both the server and the certificating body for the server. This step is not obligatory and may be omitted, if the negotiated method of key exchange does not require sending the certificate (e.g., the anonymous Diffie-Hellman method). Depending on the negotiated method of key exchange, the server may send an additional "server_key_exchange" message. This is not required in the case of fixed Diffie-Hellman method or RSA key exchange method. Moreover, the server can also request a certificate from the client. The final step of Phase 2 is the "server_done" message, which has no parameters and is sent by the server merely to indicate the end of server messages. After sending this message, the server waits for a client response.

o Client Authentication and Key Exchange

Upon receipt of the "server_done" message, the client verifies the server's certificate, the certificate validation date and path, as well as any other parameters sent by the server in the "server_hello" message. The client's verification consists of:

§ Checking the certificate validation date and its comparison with the current date, to ensure the validity of the certificate.

§ Checking whether the certifying body is included in the list of trusted Certifying Authorities of the client. If the server's CA is not included in the client's CA list, then the client attempts to verify the server CA's signature. If no information regarding the server CA can be obtained, the client terminates the identification procedure by either returning an error signal or by reporting an error message to the user.

§ Establishing the authenticity of the server CA. If the server's CA is included in the client's list of trusted CAs, then the client compares the server CA's public key with the public key available from the list of trusted CAs. A match between these two keys verifies the authenticity of the certifying body.

§ Checking whether the domain name given in server's certificate matches the server name shown in the certificate.

§ Upon successful completion of all the above steps, the server is considered authenticated. If all parameters are matched and the server's certificate correctly verified, the client sends the server one or multiple messages. The first one is the "client_key_exchange" message, which must be sent to deliver the keys. The content of this message depends on the negotiated method of key exchange. Moreover, at the server's request, the client's certificate is sent along with the message, which enables the verification of client's certificate.

o Finish or Final Step

Phase 4 is to confirm the messages so far received and to verify whether the data is correct. The client sends a "change_cipher_spec" message (in accordance with the SSL ChangeCipher Spec Protocol), after inserting the pending set of algorithm parameters and keys into it. Then the client sends a "finished" message, which confirms that the negotiated parameters and data are correct. The server in response sends the same message sequence back to the client. If the "finished" message is correctly read by both parties, then it confirms that the transmitted data, negotiated algorithms and the session key are correct. This in turn indicates that the session has been set up and that it is possible to now send application data from the server to the client, and vice versa. At this point, the TCP session between the client and the server is closed; however, a session state is maintained so that communication can be resumed within the session.

Note that phases 2 and 3 are used by both parties to verify the authenticity of the server's certificate and possibly the client's certificate. If the server cannot be successfully authenticated by the client on the basis of the delivered certificate, the handshake terminates and the client generates an error message. The same event takes place at the server end if the client's certificate cannot be authenticated.

Recommended Readings and Web References

An Introduction to Cryptography; PGP Corporation; June 8, 2004.

Cheng, P. et al. 1998. A Security Architecture for the Internet Protocol; IBM Systems Journal, Number 1; 1998.

Kent, S. and R. Atkinson, 1998. IP Encapsulating Security Payload (ESP); November 1998; IETF; RFC 2406.

MIT Kerberos Website: Bryant, W. 1998. Designing an Authentication System: A Dialogue in Four scenes; Project Athena Document; February 1998.

PGP Home Page and MIT PGP Distribution Website.

SSL/TLS in Detail; Microsoft TechNet; July, 2003.

Stallings, William. 2003. Network Security Essentials: Applications and Standards; 2nd edition; Upper Saddle River, NJ: Prentice Hall.

The SSL Protocol: Version 3.0; Netscape's final SSL 3.0 draft; November, 1996.

Part Two

Application Security: Fundamentals and Practices

Chapter 5

Application Level Attacks

An *injection* is an exploitation of an inherent weakness present in code. An attacker can either explicitly inject malicious code into the system, or (s)he can introduce junk characters as input data to force a program to deviate from its normal course of execution. Injection attacks can be easily identified by observing the behaviour of an application for a common set of inputs. This chapter discusses the not-so-good coding practices which raise serious vulnerability issues, exposing the deployed product to several types of attacks. Besides discussing the disastrous consequences of these attacks, the current chapter also familiarizes the reader with ways to overcome possible weak points of an application.

5.1 OCCURRENCES

Attacks can come from two types of interactions between a user and an application—the Data Flow and the Control flow. These two interactions have given way to several rotes of exploits, although these interactions are not the only ones responsible for such exploits. There can be other approaches as well. The exploit sources can be categorized as follows:

1. Common input vectors (data):

- Web (POST message, GET message, cookies)
- Database results
- User input (applet, standalone code)
- Data or configuration files
- Environment variables
- Network sockets
- Return values from external APIs
- Return values from external processes

2. Common output contexts (control):
   - HTML responses
   - SQL commands
   - Log files
   - Custom protocols
   - Command lines and command line arguments
   - Arguments to external APIs
   - Arguments to external processes
   - Terminal screens

5.2 CONSEQUENCES

Injection attacks are sometimes carried out for design betterment and checking the usefulness of an application. Since it is often very difficult and costly to modify code and re-deploy it, injection attacks provide a handy alternative to check inherent vulnerabilities of an application. But it must be noted that these attacks are less robust than the traditional approaches of vulnerability checking. Also, instead of being used as a beneficial tool as described above, injection attacks are mostly used as a malevolent device to exploit weaknesses in a system, thereby causing a whole host of potentially harmful and adverse effects. These effects can be summarized as follows:

1. Post junk input to increase the processing load, thereby making the system slow and causing DoS.

2. Steal authentication credentials to get unwanted access to a system and its stored information.
3. Cascade unprivileged access to illegal users and block legal accesses.
4. Install malware to disturb the normal flow of a system.
5. Modify or delete database entries.

5.3 ATTACK TYPES

Most prevalent attacks to a system fall into the following two categories:

- SQL Injection, and
- XSS.

Other common injection attacks include:

- XML Attack,
- XML Injections,
- XPATH Injections,
- Log Injection,
- LDAP Injection,
- Path Manipulation,
- HTTP Response Splitting,
- Command Injection,
- Buffer Overflow, etc.

5.4 SQL INJECTION

5.4.1 Overview

SQL Injection is a subset of Code Injection attacks where a database application fails to sanitize user input properly and is thus tricked into executing a piece of syntactically valid but malicious code. An attacker may inject special characters like the single quote ('), the equal sign (=), the comment (--) or any

SQL keyword like OR, SELECT, JOIN, UPDATE, etc., to modify the usual meaning of a query and push his/her malicious code along with.

For example, consider the following method that creates a query from user input and executes it in an SQL server:

```
public int Authenticate (string email, string
password)
{
  SqlConnection dbConn = new
  SqlConnection(dbConnStr);
  Object returnValue;

  dbConn.Open();

  SqlCommand cmd=dbConn.CreateCommand ();
  cmd.CommandText="SELECT ContactID FROM
  Person.Contact"+"WHERE
  EmailAddress = '" + email + "'AND
  PasswordHash = '"+password + "'";

  cmd.CommandType = CommandType.Text;
  returnValue = cmd.ExecuteScalar();

  return ((int)returnValue);
}
```

On a non-malicious user input ("admin@mysite.com", "pass0rd"), the above method creates the following SQL query:

```
SELECT ContactID FROM Person.Contact
WHERE EmailAddress = 'admin@mysite.com'
AND PasswordHash = 'pass0rd'
```

But an attacker may trick this method with the input ("admin@mysite.com", "' or 'a' = 'a' "). Then the generated query looks like:

```
SELECT ContactID FROM Person.Contact
WHERE EmailAddress = 'admin@mysite.com'
AND PasswordHash = '' OR 'a' = 'a'
```

This innocent-looking query silently lets the attacker enter into a system and wreak havoc thereafter.

As another example, consider the following code in Stored Procedure, which leads to the same kind of disastrous result as the previous one:

```
CREATE PROCEDURE[dbo].[Authenticate]
  @User VARCHAR(60)
  @Pass VARCHAR(60)
AS
  DECLARE @SQL varchar(255)

  SET @SQL = "Select display_name" + "FROM
  users "+ "WHERE uid = \" + @User + "'" +
  "AND pass ='"+ @Pass + "'"

  EXEC(@SQL)
GO
```

5.4.2 Consequences

1. If access rights are not set to be different for different classes of database queries like Data Manipulation Language (DML), Data Definition Language (DDL) etc., then an attacker can execute them under the same access privileges, which results in a leakage of sensitive information.
2. An attacker may use an injected query to manipulate data and enforce his own choice, e.g., changing the price list of a shopping application or changing the user login ids of any application.
3. An attacker can get more information about the database structure, its legitimate users and their respective credentials.
4. An injector can inject query in a way that makes a database drop some relations or truncate them.
5. SQL injection can also be used to grant access rights to or revoke access rights from different users.
6. It can be used to launch timing attacks.
7. By SQL injection, a database server can be overloaded with junk information, which can lead to Denial of Service condition.

5.4.3 Remediation

1. Validation of User Input

Before processing the user input, there should always be strict input validation. Several methods are available for validating input data. In an application, one or more of the following methods can be used depending on the exposure and risk involved:

- Authentication and authorization checks to determine if the user input is coming from the desired origin or not.
- Parse user input into tokens for performing Lexical analysis.
- Syntax validation to verify whether the user input follows correct sequence and format.
- Semantic analysis to verify the context of a user input.

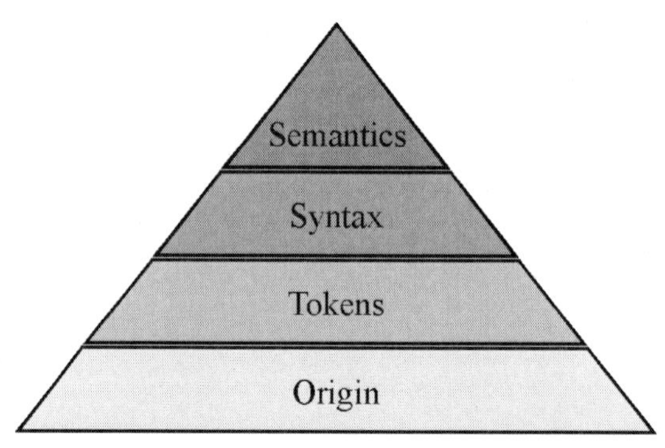

Figure 5.1: Hierarchy of Validation.

If the user input contains any unexpected character (as in Fig. 5.2), it should be rejected immediately. This is known as Black List checking.

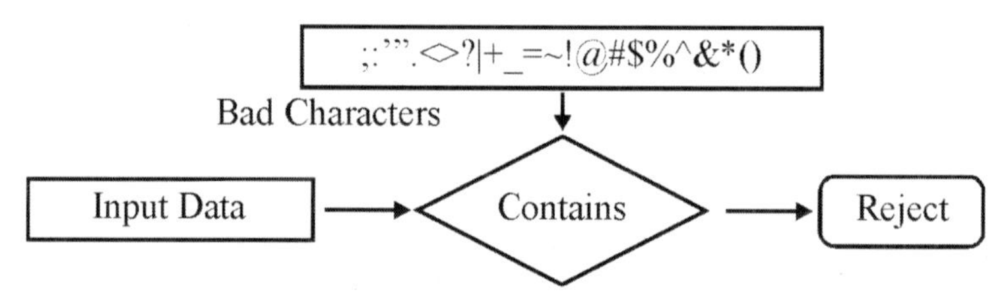

Figure 5.2: Black list checking.

After the user input passes through Black List filter, it is necessary to sanitize the input with a regular expression checker. This step, known as White List checking (Fig. 5.3), ensures syntactic consistency of the use input.

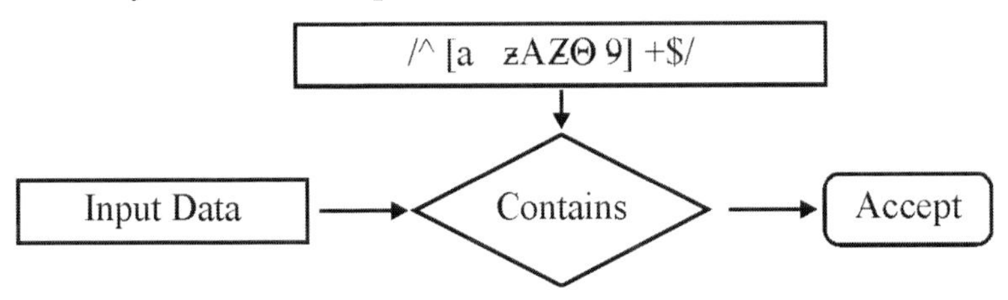

Figure 5.3: White List Checking.

2. Use the Parameters Collection with Dynamic SQL

It is safer to use the parameters collection in dynamic SQL instead of direct concatenation of commands created from the user input. Parameter placeholders in PreparedStatement automatically take care of input validation, thereby reducing the chances of injection attacks.

```
public boolean authenticate (String name, String
pass)
{
  PreparedStatement pstmt;
  String sql = "SELECT display_name from user_t
  WHERE name =? AND passwd =?";
  pstmt = this.conn.prepareStatement (sql);
  pstmt.setString (1, name);
  pstmt.setString (2, pass);
  ResultSet results = pstmt.executeQuery ();
  return results.first ();
}
```

3. Use Type-safe SQL Parameters

Although PreparedStatement reduces the fear of injection attacks somewhat, it fails to remove it completely. Therefore, it is better to use strongly typed SQL parameters that avoid default typecasting.

```
public int authenticate (string email,string
password)
{
  SqlConnection dbConn = new
  SqlConnection(dbConnStr);
  Object returnValue;

  dbConn.Open();

  SqlCommand cmd = dbConn.CreateCommand();
  cmd.CommandText = "SELECT ContactID FROM
  Person.Contact"+"WHERE
  EmailAddress = @email AND PasswordHash =
  @password";

  cmd.Parameters.Add (new SqlParameter
  ("@email", SqlDbType.VarChar,128)).Value =
  email;

  cmd.Parameters.Add (new SqlParameter
  ("@password",
  SqlDbType.VarChar,128)).Value = password;

  cmd.CommandType = CommandType.Text;

  returnValue = cmd.ExecuteScalar ();

  return ((int) returnValue);
}
```

4. Use Stored Procedures Instead of Query Execution

Instead of generating an SQL query in the front-end and executing it in the back-end by establishing a connection with the SQL server, it is better to locally store a procedure and execute it by passing parameters from application end. Thus, SP itself takes care of sanitizing the user input.

```
CREATE PROCEDURE[dbo].[Authenticate]
  @User VARCHAR(60)
  @Pass VARCHAR(60)
AS
  SELECT display_name
  FROM users
  WHERE uid = @User
  AND pass = @Pass
GO
```

5. Use placeholders in SP

Even SPs are not always safe. For example, if we create a dynamic SQL statement in a stored procedure and execute the statement inside SP, then there is a risk of SQL injection. Instead, it is advised to use placeholders in SP for passing parameters to it.

```
CREATE PROCEDURE[dbo].[Authenticate]
  @User VARCHAR(60)
  @Pass VARCHAR(60)
AS
  DECLARE @SQL varchar(255)
  DECLARE @ParamDef VARCHAR(200)

  SET @SQL = N'Select display_name FROM users
  WHERE uid = @UID AND pass = @PWD'

  SET @ ParamDef = N'@UID VARCHAR(60), @PWD
  VARCHAR(60)'

  EXECUTE sp_executesql @SQL, @ParamDef, @UID =
  @User, @PWD = @Pass
GO
```

5.5 CROSS SITE SCRIPTING (XSS)

5.5.1 Overview

Cross site scripting (XSS) is a special kind of code injection attack where the malicious code supplied by an attacker (known as a "cross-site script") gets mingled with an otherwise innocent web application. When an unsuspecting user launches this web application into his/her browser, (s)he is attacked by the cross-site script. This is possible only if the web application allows such scripting, either intentionally or unintentionally.

There are three types of XSS:

- Reflected or non-persistent XSS,
- Stored or persistent XSS,
- DOM-based XSS

5.5.1.1 Reflected cross-site scripting

In reflected XSS, the target web application has an inherent weakness: it reflects back some or all of the user input sent to the server as part of a request. For example, it may display error messages, search results, or any other response that includes some or all of the user input sent to the server. In such cases, the attacker tricks the application to fabricate the response in his/her way, so that the rendered page looks like it is coming from a trusted website, but when an unsuspecting user clicks on a link, his/her user credentials are silently purloined by a specially crafted malicious script delivered along with the webpage.

5.5.1.2 Stored cross-site scripting

Stored cross-site scripting exploits those web applications where some or the entire user data sent to a server is stored permanently in a database. This database is later used to render pages for all general users. Typical content-management software such as forums and bulletin boards where users are allowed to post raw HTML or XHTML-formatted data, are extremely prone to stored XSS. If an attacker is able to inject malicious code into such a web application, then for any subsequent unsuspecting visitor, the web application inadvertently executes malicious code along with displaying valid results.

5.5.1.3 DOM-based XSS

DOM-based XSS or Type-0 XSS is an XSS attack where under the influence of the injected code, the DOM environment in victim's browser used by original client-side script is modified, so the client-side code starts behaving abnormally. In contrast to other types of XSSs, in this attack the HTTP response is not modified, although the client-side code contained in the page executes differently due to the malicious modifications of DOM environment.

This is in contrast to other XSS attacks (stored or reflected), wherein the attack payload is placed in the response page (due to a server side flaw).

5.5.2 Consequences

1. By virtue of XSS, an attacker can capture cookies stored in local systems and thus hijack a secured session,
2. XSS can be used to fake appearance of a webpage so that it looks convincing to unsuspecting users, thereby tricking them to pass their credentials to the attacker,
3. It can be used to execute web requests masqueraded to be coming from a valid user. The generated response then silently goes to the attacker,
4. Port scanning of victim network can be achieved using XSS exploitation.

5.5.3 Remediation

1. Escape or encode all HTML and script components in the user input.
2. Do not trust client-side input validation and have adequate sanitization mechanisms in place.
3. Although difficult, it is generally recommended to store and process raw data using safe methods.

5.6 XML-RELATED ATTACKS

XML-related attacks can be broadly classified as:

- XML Entity Attacks,
- XML Injection,
- XPATH Injection.

5.6.1 XML Entity Attacks

5.6.1.1 Overview

XML entity attacks bank on some of the inherent weaknesses of popular XML parsers. They can also be triggered by ignorant and unsuspecting users without proper knowledge of DTD-related issues and by applications failing to validate data supplied in XML Body or Document Type Definitions.

5.6.1.1.1 Internal entity-based

In XML, instead of typing the same text over and over again, an internal entity is defined which contains the text. Then we only use the internal entity wherever the corresponding text needs to be inserted. The internal entity is subsequently expanded by XML parser, resulting in the underlying text being displayed in every valid location. Such an internal entity can fall prey to serious attacks, e.g., the Billion Laughs Attack.

A Billion Laughs attack is a kind of DoS attack that originally targeted XML parsers.

Applications need to parse XML in order to manipulate it, so the first thing that happens when XML hits an application is that it is parsed even before the developer can possibly check it. So, it signifies that it is the application server, not the developer which is responsible for handling this type of attack.

The following kind of messages was used to force recursive entity expansion or other repeated processing jobs that exhausted server resources and rapidly led to a Denial of Service condition.

```
<!DOCTYPE billion [
  <!ENTITY laugh1 "Ha !">
  <!ENTITY laugh2 "&laugh1; &laugh1;">
  <!ENTITY laugh3 "&laugh2; &laugh2;">
  <!ENTITY laugh4 "&laugh3; &laugh3;">
  <!ENTITY laugh5 "&laugh4; &laugh4;">
   ...
  <!ENTITY laugh128 "&laugh127; &laugh127;">
]>
<billion>& laugh128;</billion>
```

The above example results in a recursive parsing that prints millions of "Ha!" and uses up gigabytes of memory and more than 90% of CPU cycles, thereby causing DoS.

5.6.1.1.2 External entity-based

When the replacement text is too long to define as an internal entity, it is stored in a file and an external XML entity is written to point to the file. When an XML parser encounters an external entity, it looks for the corresponding file, reads its content, and finally places the replacement text in its proper position. An external entity can point to a remote URL as well. The referenced file in this case can be a normal text document or a binary file, e.g., an executable, an image or a video.

An attacker can use external entities to refer to an unauthorized file with the potential to wreak havoc. Servers can also be tricked to elicit HTTP GET requests to malicious web links. Another possibility is that the attacker refers to a huge file. When the XML parser tries to parse such a file, it gets stuck, thus causing a DoS condition.

5.6.1.2 Consequences

1. DoS using Billion Laughs Attack
2. DoS by referencing huge files or by blocking file resources
3. DoS by referencing remote URLs from where control never returns
4. Confidentiality breach by referencing unauthorized files

5. Using unauthenticated GET requests for attacking internal systems

5.6.2 XML Injection

5.6.2.1 Overview

XML injection attacks take place when XML Documents are not adequately validated against a pre-defined schema and user input is directly used (as XML fragments) to construct the final XML document using string concatenation. An attacker can use XML injection to change the meaning and structure of an XML document. It can be done by introducing new elements or overriding existing elements. A denial of service (DoS) situation can also be brought about by creating deeply nested XML fragments.

Consider an example where a user specifies the input "Apple" and it is rendered into the following XML:

```
<?xml version="1.0">
<xmlmessage>
<fruits>
    <fruit>
          Apple
    </fruit>
</fruits>
```

A malicious user can enter "Apple</fruit><fruit>Attacker Fruit" as input. This will compromise the integrity of the rendered XML, as shown below:

```
<?xml version="1.0">
<xmlmessage>
<fruits>
    <fruit>
          Apple
    </fruit>
<fruit>
          Attacker Fruit
    </fruit>
</fruits>
```

5.6.2.2 Consequences

1. XML injection can be used to remove or override valid existing elements (or nodes) of an XML document by injecting appropriate control sequences.
2. XML injection can also be used to introduce new invalid nodes, entities or attributes, thereby compromising the integrity of the resulting XML document.
3. A DoS attack can be brought about by introducing deeply nested nodes.

5.6.3 XPATH Injection

5.6.3.1 Overview

For XPATH injection, an attacker first sends semi-valid test data to a system and monitors its response. After sufficient number of responses, the attacker gleans a fairly good amount of knowledge about XPATH validation. Then (s)he starts sending data that envelopes malicious XPATH queries. This is known as XPATH injection. It may lead leakage of sensitive information and credential theft.

One principal reason behind XPath injection is an application failing to adequately sanitize user input. Note that Path injection is similar in essence to SQL Injection, but it is aimed at XML documents.

5.6.3.2 Consequences

1. XPATH injection helps an attacker gain access to user credentials or other unauthorized data.
2. XPATH injection can also be used to bypass authentication.

5.6.4 Remediation

A few general remediation strategies for XML injection are listed below:

1. Configure XML parsers to restrict the domain of DTDs that can be handled,
2. Configure a maximum entity expansion limit as well as a maximum attribute limit (to curb recursive expansion),
3. Strong user input validation and sanitization (if needed) before further processing.

5.7 LOG INJECTION

5.7.1 Overview

Log injection originates from vulnerability in the logging and tracking mechanism of an application. The attacker injects entries containing special formatting text sequence or control characters in logs and audit trails that mislead administrators in tracing the attack.

For example, consider the following code snippet that implements a logon event logger:

```
public void LogEvent (name, password)
{
    Logger l = Logger.getLogger("web");

    if (!authenticate(name, password)
  {
  l.info ("User login failed for: " + name +
  "\n");
  }
  else
  {
  l.info ("User login successful for: " + name);
  }
}
```

Two successive normal log entries for the above code look like:

User login failed for: guest

User login failed for: admin

If, however, an attacker passes "guest\nUser login successful for: admin" as username, then the above two log entries look like:

User login failed for: guest

User login successful for: admin

5.7.2 Consequences

The effect of a log injection can vary from less severe to more severe, as detailed below:

1. Log Injection corrupts the integrity of event logs and audit trails by generating fake records.
2. Log injection can be used to hide an attack, e.g., by pushing some line feed or backspace characters into the log.
3. Log injection can also be used to bring down the log monitoring and display system.

5.7.3 Remediation

1. Adding sequence number for each log entry reduces the chances of entering fake or additional log entries.
2. Before adding to a log, a log entry must be validated by a white list regular expression filter. This helps avoid special characters and command sequences in log entries. However, handling of false positives and false negatives remains a crucial issue. Log experts and system administrators should come together to resolve it.

5.8 PATH MANIPULATION

5.8.1 Overview

A path manipulation attack occurs when the attacker gains access to the location of an unprivileged file by tricking an application to accept unwanted path specifications.

5.8.2 Consequences

1. By manipulating the path at will, an attacker can access and control most physical locations in a system, including the unprivileged ones.
2. Path manipulation may ultimately lead to corruption or overwrites of important files in a system. The consequences can be devastating.

5.8.3 Remediation

1. Before doing any operation on user input, an application must verify the absolute file address specified in it.
2. Special path separating characters or any such sequence must not be allowed as part of the user input. Again, false positives and false negatives must be taken care of.

5.9 HTTP RESPONSE SPLITTING

5.9.1 Overview

HTTP response splitting is a form of web application vulnerability where malicious input from attacker gets inserted into HTTP response header and sent to user which results into attacker gaining complete access on both HTTP response header and body. While the attacker gains control over header by inserting unwanted sequence of line feeds and carriage returns, it becomes easier for him to get access not only the response body but also it allows him to split the response into two or to create additional responses altogether.

5.9.2 Consequences

1. HTTP response splitting may lead to XSS attacks, cross-user defacement, web cache poisoning and many other related problems.

2. The HTTP response loses its integrity as a result of the attack.

5.9.3 Remediation

1. User input must be adequately sanitized to prevent unwanted control sequences in the HTTP response header.
2. Before sending the response to a user, it should be validated for malicious characters.

5.10 LDAP INJECTION

5.10.1 Overview

Lightweight Directory Access protocol (LDAP) injection is a vulnerability of web applications where the application executes LDAP statements using local proxy based on inadequately sanitized user input. This may lead to abnormal behaviour of LDAP query execution or execution of unwanted queries.

5.10.2 Consequences

1. LDAP injection may lead to content modification in the LDAP tree.
2. By LDAP injection, an attacker may force a system to run unwanted queries and thereby granting access permissions to unwanted users.

5.10.3 Remediation

1. User must be adequately validated using a white list check and sanitized if possible.
2. Code execution permissions can be altered systematically to keep a system from executing unwanted queries.

5.11 COMMAND INJECTION

5.11.1 Overview

Command injection attacks result from a type of vulnerability where an application takes text input from users as part of a command and executes the command under its own execution privileges. So, if an attacker passes properly delimited extra commands along with the valid ones to the application, then those extra commands also get executed no matter what access rights the attacker has. This leads to many problems, as detailed next.

5.11.2 Consequences

1. Since the injected commands run under the application's execution privileges, the former can be used to extend execution permissions to unauthorized users and get more information about the current system.
2. The injected commands may harm the system directly by corrupting system memory or crashing system processes.

5.11.3 Remediation

1. Adequate validation and sanitization (if possible) of user input.
2. System commands like exec(), system(), etc., must be avoided in an application as far as possible. Raw text command or file information from user input must not be used. Special APIs should be used to run commands under appropriate protection and access level security.

5.12 BUFFER OVERFLOW

5.12.1 Overview

A buffer overflow, or buffer overrun, is an anomaly where a program tries to put more data into memory than what is allocated to it by the system. This causes overrunning of current buffer's boundary and a possible overwrite into adjacent memory.

5.12.2 Consequences

1. Buffer overflow may lead to loss of data due to memory overwrite.
2. It may also lead to memory *leakage* and crash and memory access errors.
3. It can cause erratic behaviour of the current program or other programs using nearby memory slots.

5.12.3 Remediation

1. Use of a programming language that comes with default boundary checking.
2. If default boundary checks are not available, then manual or semi-automatic checks for input length are recommended. Special APIs that protect memory overwrite can also be used.

5.13 CROSS-SITE REQUEST FORGERY (CSRF)

5.13.1 Overview

Cross-site request forgery, one-click attack or session riding is a type of malicious exploit of a web application that aims at destroying the trust a user has for the particular website. It works through user interface interactions using GET or POST messages. The attacker hides a malicious link under some text or image. When a user unknowingly clicks the link while having an active authenticated session with the trusted site, the authentication information gets posted to the link embedded by the attacker. This is why the attack is known as "one-click attack".

5.13.2 Consequences

1. CSRF may be used to steal user authentication details, profile information, money or email addresses.
2. It can lead to change in security information of victim.

5.13.3 Remediation

1. Ensuring that an HTTP request is coming legitimately from the application's user interface and not from any other source.
2. By adding a hidden form field to a web form, user identification can be validated.
3. The CSRF prevention token should be random and dynamically generated by application runtime, so that it becomes difficult for an attacker to guess.

Recommended Readings and Web References

ha.ckers.org: XSS (Cross Site Scripting) Cheat Sheet.

MSDN Magazine: Stop SQL Injection Attacks Before They Stop You.

MSDN: How To: Protect From Injection Attacks in ASP.NET.

MSDN: How To: Protect From SQL Injection in ASP.NET.

Open Web Application Security Project Website: Different threats and vulnerabilities.

SQLsecurity.com: SQL Injection Attacks.

CHAPTER 6

Practical Software Security—ASP.Net and Java

Learning about security threats involved in an application helps in taking preventive measures and building security appliances. But note that most of the threats faced by an application are due to one or more of its inherent weaknesses. So, it is recommended to follow best coding practices while developing a modern software application. Present-day software development tools and programming frameworks usually come with better language support with in-built security mechanisms. Developers and software professionals should therefore be aware of the available security features of different programming frameworks. This chapter discusses security guidelines prescribed by two most popular development frameworks—ASP.Net and Java.

6.1 ASP.NET SECURITY GUIDELINES

6.1.1 Overview

ASP.Net web development is based on a wide range of language support mainly like C#, VB.Net, etc. The .Net framework operates on a Common Language Runtime (CLR) that interacts with underlying operating systems. The supported languages communicate with each other and the CLR by a secured communication mechanism provided by .Net code security and type safety policies. They all use a common set of dedicated libraries to make system calls via the framework. It also comes with memory management, automatic bound checking and restricted pointer support with prominent distinction with type safe managed codes. The .Net framework also introduced some built-in security mechanisms. Some of their important features are described in the following sections.

6.1.2 Code Access Security (CAS)

In the age of internet, users and user applications are always exposed to unknown threats arising from unintended execution of codes from unknown sources. Even when all security mechanisms are in place, some of these codes can still be executed owing to the fact that they have the same access privileges as that of the application which is (unknowingly) working on behalf of them.

To prevent execution of unknown codes in an application, the .Net framework provides a unique option: the Code Access Security (CAS). Every application that targets the framework must run on the Common Language Runtime (CLR) for interacting with the underlying Operating System or other collaborating applications. The CLR evaluates the code and applies sufficient level of security and permissions based on the nature of the code. Note, however, that it is not always possible to be sure that the CLR will assign an accurate level of permission for the code.

The Code Access Security (CAS) is a model for enforcing security as defined by the .Net framework. It restricts the type of resources a particular code can access and the type of operations it can perform on the allowed resources. This restriction is independent of user roles and user security privileges.

Using CAS, the trust level of code can be configured using the following configuration setting:

```
<trust level="Full|High|Medium|Low|Minimal" />
```

6.1.3 Windows CardSpace

In today's world with round-the-globe online accessibility, it is very important for applications and systems to judge the identity of users before allowing them to access legitimate resources. "Identity", however, is difficult to define as users spread out across different countries and different demographics using very different hardware and software platforms. So, in such a heterogeneous environment, "identity" cannot be defined by a single standard and must come as a conglomeration of different facets. Microsoft in this regard plays a vital role to establish a common meta-system for identity management, where users using different hardware devices and software platforms can easily and effectively use their own local digital identity. Windows CardSpace is such a technology that defines an identity meta-system for multiple varieties of digital identity.

A *digital identity* is a security token defined by a system and assigned to a user. A set of these identities of various users is maintained in a database (e.g., Active Directory). A security token is a byte array that contains user information like username, X.509 user authentication certificate, etc. Windows CardSpace addresses the challenge to maintain and manage the wide variety of digital identities by designing an Identity Meta-system. It provides improved user confidence in the identity of remote applications and vice versa by removing standard password-based user login mechanism and supporting consistent and coherent use of different meta-systems.

Windows CardSpace uses information cards—visual representation of a user's digital identity and technically an XML document stored in the user's local Windows system. These cards enable the user to communicate with his/her identity provider and request for security tokens for service-related communications. The cards also help CardSpace to verify and match certificates with relying parties before communication begins. Each information card contains a globally unique CardSpace Reference Identifier, URLs of service policies, URLs where security tokens can be requested, types of security tokens, timestamp of issue and an image for visual identity. Using these information cards, Windows CardSpace effectively generates a security framework for validating user identity. It allows users to interact with service parties, share security tokens and certificates with relying end-points and have an easy way of communication in the internet.

6.1.4 MachineKey Configuration

Most applications use hidden form fields for maintaining sensitive information. Information contained in hidden fields is generally stored as ViewStates in the local machine of users. If an attacker gets access to the ViewState data, then they can change behaviour of forms or get access to user information by tampering ViewState data. <machineKey> is a standard element defined in Web.config file of ASP.Net application, which under appropriate configurations can be used to control tamper proofing and encryption of ViewState, form authentication tickets and role cookies. By default ViewStates are signed and tamper-proof, but the default behaviour may need to be altered if the web application is running in a Web farm or the user needs to share authentication tickets across applications.

The default settings for <pages> and <machineKey> elements are defined in the machine-level web.config.comments file. The relevant default settings are as follows:

```
<pages enableViewStateMac="true"
viewStateEncryptionMode="Auto" ... />

<machineKey validationKey="AutoGenerate,IsolateApps"
        decryptionKey="AutoGenerate,IsolateApps"
        validation="SHA1" decryption="Auto" />
```

If the enableViewStateMac attribute of <page> element contains true value, then the <machineKey> element is investigated. The "validation" attribute specifies a hashing algorithm and there are two keys for this authentication mechanism. Based on values of attributes, the hash function generates a Message Authentication Code (MAC) from ViewState content and this value is compared in subsequent requests to keep ViewState information from being tampered. The value of validationKey is used to make ViewState tamper-proof and it is also used to sign authentication tickets for forms authentication; whereas the value of decryptionKey is used to encrypt and decrypt authentication tickets.

6.1.5 Authentication in .Net

6.1.5.1 Relation between ASP.Net and IIS authentication

Security is always the major concern for both web developers and deployment architects as the application which deals with sensitive user information is always prone to attacks from malicious users. Thus for a developer it is of utmost necessity to understand the security models provided by the development framework and the deployment platform. IIS serves as a scalable, fast and easy-to-deploy environment for hosting web applications. It provides many authentication mechanisms to provide security for hosted web sites. Being one of the most popular development platforms, ASP.Net is also equipped with several different security and authentication mechanisms of its own.

As a developer, it is thus necessary to understand the relation between security mechanisms of ASP.Net and those of IIS. A developer is expected to know different combinations of security procedures of IIS and those of ASP.Net that ensures better scalability, performance and security of the concerned web application. IIS maintains security-related configurations and information in the IIS metabase, whereas ASP.Net maintains the same security settings and configurations in XML files. The following diagram (Fig. 6.1) shows the most likely combinations between IIS and ASP.NET.

6.1.5.2 ASP.Net authentication

Apart from the IIS authentication, ASP.Net supports additional authentication facilities using authentication providers like:

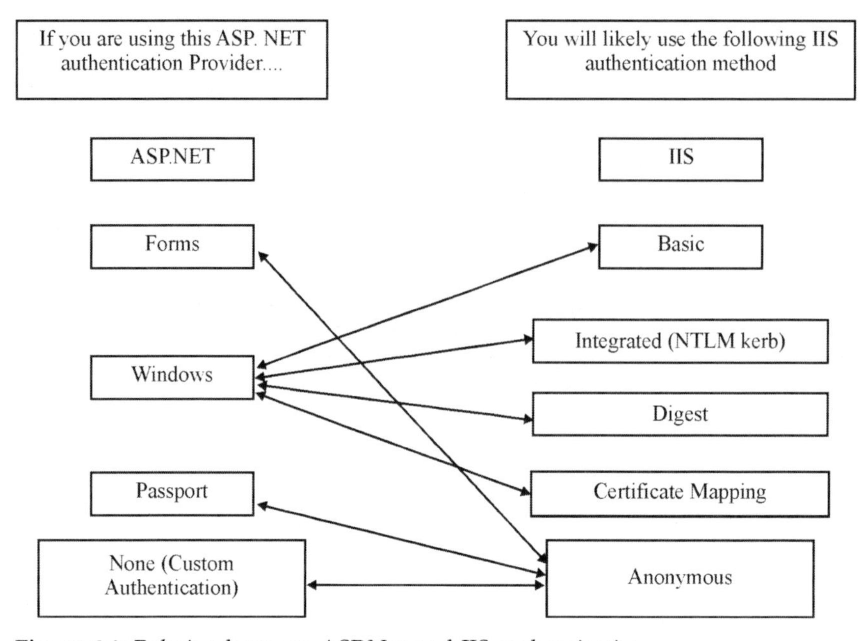

Figure 6.1: Relation between ASP.Net and IIS authentication.

- Windows (default)
- Forms
- Passport
- None (Custom Authentication)

To enable an authentication provider for an ASP.NET application, the authentication element in either machine.config or Web.config can be configured as follows:

```
<system.web>
   <!-- mode=[Windows|Forms|Passport|None] -->
   <authentication mode="Windows" />
</system.web>
```

Each ASP.NET authentication provider supports an OnAuthenticate event that occurs during the authentication process, which can be used to implement a custom authorization scheme. The primary purpose of this event is to attach to the context a custom object that implements the IPrincipal Interface.

Whatever may be the provider configured for ASP.Net authentication, it will only be in action only after the execution of IIS authentication. IIS sends an authentication token after successful authentication and depending on this token, ASP.Net proceeds further.

- ***Forms (Cookie)***

In Forms Authentication scheme, ASP.Net enables developers to design a custom login page with the desired look-and-feel, which can be prompted to clients to provide their credentials. On submission of this login form, the web server checks these credentials against a standard database detailing information of all authenticated users. Once an exact match is found, the server creates a legitimate session and assigns a cookie to the client which eventually gets stored in the client workstation. On subsequent requests, the client sends this cookie to the server to authenticate it.

This authentication provider gives developer the flexibility of designing a custom login page. As the credentials are verified against the database, the application does not need to keep track of individual windows accounts. But a major vulnerability of this scheme is related to client-side trust on cookies. The cookies are always prone to replay attacks. If an attacker gets access to a legitimate cookie, (s)he can easily masquerade himself/herself as a valid user.

To implement forms authentication, the Web.config file can be configured as follows:

```
<!-- Web.config file -->
<system.web>
   <authentication mode="Forms">
     <forms forms="401kApp" loginUrl="/login.aspx"
/>
   </authentication>
</system.web>
```

It is a general norm of implementation of Anonymous IIS authentication along with Forms authentication of IIS.

- *Windows*

It strictly behoves the IIS to perform the required authentication for a client. After IIS authenticates a client, it passes a security token to ASP.NET. ASP.NET then constructs and attaches an object of the WindowsPrincipal Class to the application context based on the security token it received from IIS.

Although it reduces the developer's burden by lessening the need for writing custom authentication mechanisms, but as IIS authenticates users based on Windows user accounts, the bottleneck now shifts to a rigorous maintenance of Windows Account.

- *Passport*

The Passport authentication provider is a forms-based, centralized authentication service which offers a single logon supported by most modern-day browsers and core profile services for member sites. On registration of a member site with the central service, it grants a site-specific key. This key is used further to encrypt and decrypt the query strings passed between the member site and the Passport logon server. This is somewhat limited in the sense that it depends on an external agent for authentication.

To implement Passport, the Passport SDK must be installed first. Then the Web.config file needs to be configured as follows:

```
<!-- Web.config file -->
<system.web>
   <authentication mode="Passport" />
</system.web>
```

- *None (Custom Authentication)*

This authentication service is implemented when there is absolutely no requirement for user authentication before granting access to web resources. This can be the case where sites have open display of contents or are meant for public advertisement.

It can also be used if the developer does not want to depend on standard authentication schemes and wants to write a custom implementation from scratch. It can be achieved using an ISAPI filter that authenticates users and manually creates an object of the GenericPrincipal Class. Thus in such cases it provides full control to developers on handling authentication, but this way it also increases the overhead of development.

To implement no authentication for developing custom authentication, a custom ISAPI filter to bypass IIS authentication has to be designed and the following Web.config configuration needs to be incorporated:

```
<!-- Web.config file -->
<system.web>
   <authentication mode="None" />
</system.web>
```

6.1.5.3 IIS authentication

It is of critical importance for modern-day web applications to validate and authenticate users before granting access to distributed web resources. Even when application-level authentications are in place, it may sometimes become necessary for web servers to provide another layer of authentication. Like other popular servers, IIS also comes with a few authentication schemes of its own to verify the identity of a user.

- ***Basic***

The basic authentication in IIS, which is part of the HTTP 1.0 specification, uses Windows user accounts. In this scheme, the user is prompted to provide username and credentials. This data is sent to the server in Base64 encoding. The server decodes the credentials and validates the authenticity of the user.

As Base64 is involved, it is relatively easy for a third party to steal the credentials and decode them. Thus, the basic scheme is not that effective, although being part of a standard is widely supported by all browsers. For improving the security of credentials exchange, SSL/TLS can be used to establish a secured session between the communicating parties. This scheme can further be extended using Kerberos and X.509 certificates.

- ***Integrated Windows Authentication (IWS)***

Integrated Windows authentication was formerly known as NT LAN Manager (NTLM) authentication and Windows NT Challenge/Response authentication. It is supported by latest browsers like Internet Explorer 2.0 or later, running on server platforms like IIS 5.0 or later and using operating system frameworks like Windows Server 2000.

The procedure of authentication is based on an exchange of two authentication headers—Negotiate and NTLM. Negotiate is the default header when all participating entities run on the same Windows platform. For all other cases, NTLM is chosen. The choice of a header is usually based on a negotiation between the browser and IIS.

Unlike the basic scheme, IWS sends user credentials across the network as a hash (e.g., MD5). It is by far the strongest authentication scheme and gives the highest degree of performance when operated on Windows accounts in conjunction with Kerberos v5. Coupled with Kerberos v5 authentication, IIS can securely distribute user credentials among computers running Windows 2000 and later that have been properly certified and configured for delegation. Delegation enables remote access of resources on behalf of the delegated user. The only limitation of this scheme is that it is highly dependent on latest Windows-based server technologies and browser support.

- ***Digest***

The fundamental weakness of the basic authentication scheme stems from the exchange of user credentials like passwords over the internet in Base64-encoded plain-text format. Digest is introduced to address this issue. Instead of plain-text message, Digest generates a fixed-length hash using standard hashing schemes like SHA-1, MD5, etc. This hash value is known as a Message Digest.

When a client requests a service, the server asks for this digest from the client. The client then attaches his/her password and data and signs the data with a common key known to both client and server, and sends it along to the server. The server then accesses Active Directory to get the public copy of client password and prepares a digest of the known data using the same hashing mechanism. It then compares its generated digest and the client's digest. On success, the server grants access of service to the client.

It is a stronger scheme than Basic, as an encrypted digest is shared instead of (almost) raw passwords. But this scheme is also limited in the sense that it cannot delegate credentials at the same time. Another issue comes from replay attacks on the digest. It can be mitigated using SSL. Digest can work in tandem with proxy servers and firewalls.

- ***Client Certificate Mapping (CCM)***

This scheme is inherently compatible with the Windows authentication provider of ASP.Net. CCM relies on third parties for authentication. Usually, Certification Authorities (CA) who issue public key certificates like X.509 certificate are involved in between. But in CCM, end points and service providers directly interact with each other based on public key certificates and service session keys.

CCM provides a strong authentication scheme, but the configuration procedure is browser-specific and rather cumbersome to manage.

- ***Anonymous***

Although listed as an authentication scheme, "Anonymous" is not an authentication scheme per se, because it does not prompt users for credentials, and directly grants them access to network resources. As there is no overhead of a verification mechanism, this scheme is the fastest, thereby producing the highest yield. This is most effective for managing access to public areas of websites. As there is no authentication and credentials verification, IIS must rely on local logon of the client workstation for them.

6.1.6 Restricting Configuration Override

There are different configuration files, e.g., Machine.config, ApplicationName.config, Web.config, etc. These different configuration files have different configuration settings. The .Net development platform defines an inheritance hierarchy on these settings. In most configuration files, developers define security settings, redirection URLs, authentication and encryption policies, etc. An attacker can lock such settings by overriding the configuration files. The code that prevents such configuration overrides is given below:

```
<configuration>
<location path="application1"
allowOverride="false">
    <system.web>
          <trust level = "Medium" />
          </system.web>
</location>
</configuration>
```

6.2 JAVA SECURITY GUIDELINES

6.2.1 Java Security Model

The Java framework is one of the most widely used software frameworks for web development. Java gained popularity among developers, mainly due to its large number of libraries supporting a wide range of features and also because many of the software written in Java are open-source. Being a modular language with clean documentation, Java has been very appealing to developers in general. But note that its ease of coding has also made it vulnerable towards many types of attacks.

And this is why the Java development team came up with a language-integrated security model that will give protection against such vulnerabilities. The Java security model is based on three major components of Java framework—the Compiler, the Class Loader and the Security Manager. Its architecture is explained in the following diagram (Fig. 6.2).

The original security model provided by the Java platform is known as the "sandbox model". It provides a restricted environment that runs unreliable code gleaned from the open network. The essence of the sandbox model is that local code is trusted to have full access to vital system resources (e.g., the file system, network connections, etc.), whereas downloaded

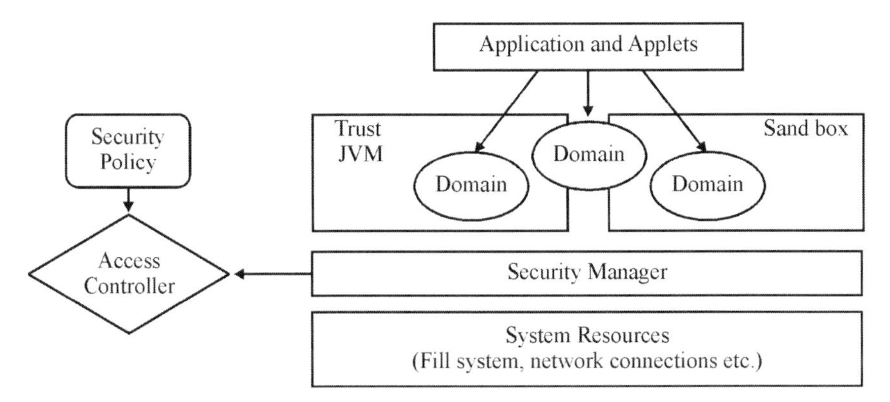

Figure 6.2: Java Security Model.

(remote) code and scripts (e.g., an applet) are not trusted and can only access the limited resources provided within a sandbox.

The Java language has overcome many limitations of earlier languages by introducing a garbage collector, a memory management mechanism and an automatic bound-checking procedure. These mechanisms help minimize the development glitches and errors. In Java, the compiler and the class loader work on top of these mechanisms and along with their routine tasks, they play a key role in ensuring *code security*. Complier and byte-code verifier ensure the execution of legitimate byte codes, and the Java Virtual Machine (JVM) guarantees type-safety at runtime. By demanding local namespaces, the class loader ensures that no outside code like applets or remote scripts can meddle with the execution of inside applications. In Java security model, the JVM is used to execute local code as highly-trusted and remote code as untrusted (i.e., the remote code goes to a sandbox, which is very restrictive).

6.2.2 Specifying Security Constraints

Web application developers often need to analyze the developed code and specify security constraints before its deployment. These security constraints can either be setup programmatically by security annotations (as described below) or declaratively as in deployment descriptors of a configuration file.

Popular security Annotations are:

- The @DeclareRoles annotation. It is used to define *security roles* in accordance with the security model of the web application. This annotation usually annotates a class. The general syntax is given below:

```
@DeclareRoles("BusinessAdmin")
public class CalculatorServlet {
//...
}

Or

@DeclareRoles({"Administrator", "Manager",
"Employee"}) for multiple roles.
```

- The @RunAs annotation. It defines the security role of an application during its execution in a Java Enterprise Edition container. It is also specified on a class, allowing developers to execute an application under a particular security role. The syntax is similar to that of the @DeclareRoles annotation, as given below:

```
@RunAs("Admin")
public class CalculatorServlet {
        //...
}
```

In the declarative style of security description, *deployment descriptors* are specified in the Web.xml file. A standard example is as follows:

```
<?xml version="1.0" encoding="ISO-8859-1"?>
<web-app xmlns="http://java.sun.com/xml/ns/j2ee"
          xmlns:xsi="http://www.w3.org/2001/
   XMLSchema-instance"

  xsi:schemaLocation="http://java.sun.com/xml/ns/j2ee
          http://java.sun.com/xml/ns/j2ee/web-
   app_2_5.xsd"
             version="2.5">
    <display-name>A Secure Application</display-
name>

    <!-- SERVLET -->
```

```
    <servlet>
          <servlet-name>catalog</servlet-name>
          <servlet-
class>com.mycorp.CatalogServlet</servlet-class>
          <init-param>
                <param-name>catalog</param-name>
                <param-value>Spring</param-value>
           </init-param>
           <security-role-ref>
                 <role-name>MGR</role-name>
                 <!-- role name used in code -->
                 <role-link>manager</role-link>
           </security-role-ref>
    </servlet>

    <!-- SECURITY ROLE -->
    <security-role>
          <role-name>manager</role-name>
    </security-role>

    <servlet-mapping>
          <servlet-name>catalog</servlet-name>
          <url-pattern>/catalog/*</url-pattern>
    </servlet-mapping>
    <!-- SECURITY CONSTRAINT -->
    <security-constraint>
          <web-resource-collection>
                <web-resource-name>CartInfo</web-
resource-name>
                <url-pattern>/catalog/cart/*</url-pattern>
                <http-method>GET</http-method>
                <http-method>POST</http-method>
          </web-resource-collection>
          <auth-constraint>
                <role-name>manager</role-name>
          </auth-constraint>
          <user-data-constraint>
                <transport-
guarantee>CONFIDENTIAL</transport-guarantee>
          </user-data-constraint>
    </security-constraint>

     <!-- LOGIN CONFIGURATION-->
    <login-config>
          <auth-method>BASIC</auth-method>
    </login-config>
</web-app>
```

The tags appearing under <web-app> serve several different purposes, as described below:

- <security-role-ref> is a descriptive tag that acts as a mapping between role-names and role-links. A role-name is the name of user security role as specified in the code. A role-link is one of the security roles defined in the <security-role> element.
- <security-role> specifies a set of abstract names for security roles used by an application. The roles help define the accessibility of different web resources.
- The <security-constraint> element is used to define a mapping between authorization constraints, web resources and network security requirements. It has several sub-elements, as described below:
  - o The <web-resource-collection> element defines a logical name for a web resource, the URL pattern for such resources and HTTP methods involved to access those resources.
  - o The <auth-constraint> element declares the security role-name for an application. This role-name specifies the web-resource-collection allowed to be accessible by the application.
  - o The <user-data-constraint> element defines how the user data will be protected in transit between the client and the container. If <transport-guarantee> is specified as INTEGRAL or CONFIDENTIAL, then all user credentials are communicated across a secure channel (using HTTPS).
- The <login-config> element is used to specify user authentication method to be used for accessing web content. User authentication can be achieved in several different ways depending on the value of <auth-method> element. It can be BASIC, FORM or CLIENT-CERT. These authentication methods are briefly described below:
  - o **HTTP Basic Authentication:** In this approach, the web server requests username and password from the client by prompting a standard dialog box and then declares

the user as an authorized user based on the comparison performed against a database containing information about all authorized users.

The syntax is as follows:

```
<login-config>
   <auth-method>BASIC</auth-method>
</login-config>
```

- o **Form-based Authentication:** The form-based authentication has two extra advantages—flexibility of development of a customized login screen and error pages with the desired look and feel. The customized login screen is displayed to the client to provide credentials. On submission of this form, the supplied information is verified to authenticate the user.

The standard syntax is as follows:

```
<login-config>
   <auth-method>FORM</auth-method>
   <realm-name>file</realm-name>
   <form-login-config>
         <form-login-page>/logon.jsp</form-
  login-page>
         <form-error-page>/logonError.jsp</
  form-error-page>
    </form-login-config>
</login-config>
```

- o **HTTPS Client Authentication:** It involves the use of Public Key Certificates like Kerberos or X.509. In this mode, the credentials exchange is performed by HTTPS (HTTP over SSL). The corresponding value for <auth-method> element is "CLIENT-CERT".

Recommended Readings and Web References

Jason Brittain and Ian F. Darwin. 2007. Tomcat: The Definitive Guide; O'Reilly Media; Second Edition; October 2007.

JavaWorld: Secure a Web application, Java-style.
MSDN: ASP.NET Authentication.
MSDN: Authentication in ASP.NET: .NET Security Guidance.
MSDN: Code Access Security.
MSDN: How To: Configure MachineKey in ASP.NET 2.0.
MSDN: IIS Authentication.
MSDN: Introducing Windows CardSpace.
MSDN: Security Guidelines: ASP.NET 2.0.
The Java EE 5 Tutorial: Securing Web Application.

CHAPTER 7

Securing Some Application—Specific Networks

In this era of ever-emerging technological advances, network-related applications have grown in quality and quantity by leaps and bounds. But this momentous growth could not eliminate the lurking fear of security breaches that continued growing as well (in sophistication and quantity). It therefore is of paramount importance to adopt specialized security measures for sophisticated technologies. This chapter talks about two such emerging technologies—SAN and VoIP, the risks involved in them and the security measures to overcome them.

7.1 SECURING STORAGE AREA NETWORKS

7.1.1 Overview

Due to the recent growth in e-commerce and the internet, the IT world has witnessed a humongous explosion of data. The global IT storage market is one of the fastest-growing fields, and requires constant and careful attention from managers and system administrators. More and more organizations are trying to focus in this field, as their image to the users and in the market is heavily dependent on the policy they undertake to manage this huge amount of data. It has been a challenge for several companies to maintain the highest level of system uptime and data availability 24 hours a day and seven days a week. Due to immense competition in the market to grab customers and keep them, companies all too often ask IT for a better technology to store and manage databases. With demands rising more than ever, two newer technologies—Network-attached Storage (NAS) and Storage Area Network (SAN), have come into play. While NAS mostly deals with the low-end market, SAN is aimed at the high-end market. SAN handles the situation by separating data from the server's operation, thereby reducing effective server load. SAN dedicates a sophisticated specialized network for data storage, management and continuous monitoring of data.

7.1.2 Purpose behind SAN

SAN comes as a solution to the constant market demand for more storage—storage with greater capacity, better security, higher efficiency (throughput) and higher reliability. SAN addressed the following requirements put up by industries:

- Meet certain expectations as part of business transformations while understanding requirements throughout the enterprise
- A scalable design that is flexible enough to accommodate future needs and enhancements without disruption
- An extremely resilient system that incorporates fault tolerance using concepts like redundancy and mirroring
- A simplified and fast fault monitoring, diagnostics and recovery system that needs minimal human intervention
- A strong and trustworthy backup system

To ensure that SAN meets the given requirements, it comes with the following characteristics:

- Highly scalable components with hot-plugging capabilities, i.e., non-disruptive server and storage maintenance
- Integration capability with high-end security systems and software

- Intelligent routing and rerouting
- Dynamic failover protection
- Hardware zoning for creating safe and secure environments
- Predictive fabric management
- Auto-configuration
- Built-in redundancy and mirroring

7.1.3 SAN Design Components

A SAN consists of:

1. Fibre Channel Switches (also called "SAN Fabric"),
2. SAN Fabric Management and Monitoring Software,
3. SAN Fabric Security and Access Control Software,
4. Storage Devices,
5. Hosts and Host Bus Adapters (HBA),
6. Cabling and Cable Connectors,
7. Gigabit Interface Converters (GBICs) that convert optical signals to electrical signals.

7.1.4 SAN Security Issues

Insulating a SAN from its external environment is not always sufficient for securing its integrity, especially when many departments of a company share different parts of the SAN data. If users with limited access permission can somehow access data of some other part of SAN, be it purposely or accidentally, then the security of the whole SAN is compromised.

Two major vulnerabilities experienced in a SAN are:

- Man-in-the-middle attacks
- SNMP vulnerabilities

Integrity of SAN is compromised if unintended and unauthorized individuals have access to certain elements of SAN management. Some of the inappropriate accesses to SAN configurations are:

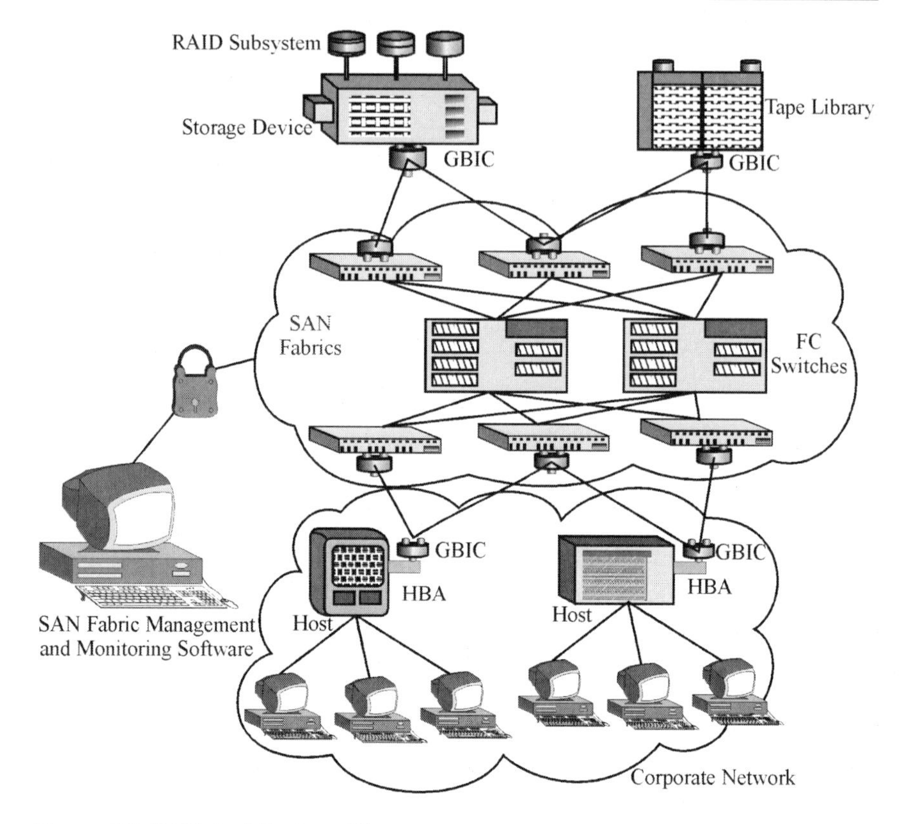

Figure 7.1: SAN Architecture Diagram.

- Exposed network administration passwords allowing unauthorized individuals to access SAN in the role of an administrator.
- Changes to zoning information allowing access to storage and read/write of data.
- Changes to security and access control policies allowing unauthorized servers or switches gain access to SAN.

7.1.4.1 Man-in-the-middle attacks

Man-in-the-middle attack is a form of (active) eavesdropping where an attacker places himself in between an active sender and an active receiver by creating two independent connections with them. The attacker is thus able to intercept all messages travelling from the sender to the receiver. The attacker can either relay the messages unchanged, or (s)he can append/alter content to create a new message. This is one of the attacks that originated from the insiders of a company. Since they are legitimate users, they can easily access security information and exploit architectural flaws of a system.

There are several possible man-in-the-middle attacks to a SAN:

1. **World Wide Name (WWN) attacks on the HBA:** The World Wide Name (WWN) attack occurs when a subsystem with different HBA and WWN ID gets access to an unauthorized portion of storage resources. It leads to compromise in confidentiality, availability and integrity of the data. One way to launch this attack is by using a compromised dual-home host with a Host Bus Adapter (HBA) that reads, stores, and distributes SAN files.
2. **Management Admin Attack:** A SAN management attack occurs when an unauthorized individual in the network decrypts the administrator password (e.g., via telnet service) and gains control over the whole SAN management subsystem.

7.1.4.2 SNMP vulnerabilities

The Simple Network Management Protocol (SNMP) is a widely deployed and standard protocol for monitoring and managing network devices. From the very beginning of SNMP, it has been used by SAN systems. SNMP allows network devices to communicate information regarding their operational states to a central system. SNMP is vulnerable to the Denial of Service (DoS) condition, service interruptions and in some cases, an attacker gaining access to an affected device. All three can seriously compromise the integrity, availability and confidentiality of SAN fabric and the data being stored. Also, the way SNMP messages travel between SAN endpoints and the processes involved in encoding and decoding them are inherently susceptible to different types of attacks that may lead to compromise of confidentiality, integrity and availability of SAN data.

7.1.5 Security Measures for SAN

1. Elements of administrative communication such as passwords must be secured on some interfaces between the security management function and the switch fabric. Since administrator-level fabric password provides the primary control over security configurations, securing these passwords is of special importance.
2. It is important to make sure that the users are only accessing and aware of those files they have been given access to, and not the others that are also lying on the same storage devices. It can be achieved by masking off the Logical Units (LUNs) that are not legitimately available to a user.
3. Hardware zoning of Servers and LUNs through Fiber Channel Switch along with port-level masking can be used to ensure security. One or more of the switches can act as trusted devices in charge of zoning change and other security-related functions.
4. HBAs can be masked using HBA drivers with masking utility that uses the WWNs supplied with them.
5. Device Connection Controls can be used to bind a particular WWN to a specific switch port or a set of ports for preventing ports in another physical location from assuming the identity of an actual WWN.
6. High-end SAN management software can be used to encrypt passwords from some interfaces (e.g., the Management Console) to a switch fabric.
7. The Management Console can be placed in an isolated, dedicated network to protect it from man-in-the-middle attack.
8. The Public Key Infrastructure (PKI) technology can be used to support the most comprehensive security solution for SAN environments by providing Access Control Locks and

Digital Certification for specific switches and WWNs to specific ports.

7.2 SECURING VOIP-ENABLED NETWORKS

7.2.1 Overview

Voice over Internet Protocol (VoIP) is the protocol that defines the ability to make calls, carry out video-conferences and send faxes over IP-based networks. VoIP has been very popular since its introduction in the field of telecommunications. We can even venture to say that the traditional Public Switched Telephone Network (PSTN) is going to be replaced by its most promising alternative VoIP in near future. Today's people prefer to make calls with IP phones or application programs like Google Talk and Skype. Many telecommunications companies and other organizations have been moving their telephony infrastructure to their data networks, since VoIP is more efficient in utilizing the available bandwidth and also presents increased flexibility and reduced management overhead. Although VoIP is already in wide use as an alternative telephony system, it is still a developing technology. And like other IT products, it also comes with its own inherent risks that need attention and protective measures.

7.2.2 Why VoIP?

The reason VOIP is so popular is that it gives significant benefits as compared to the legacy phone system. The key benefits are as follows:

1) **Cost Savings:** The most attractive feature of VoIP is that the call charge is miniscule as compared to the legacy PSTN. VoIP takes advantage of existing WAN connectivity to remote locations over a dedicated data network or the internet, thereby reducing long-distance toll charges. It is based on software rather than hardware; therefore the recurring maintenance charge is also minimal. Furthermore,

deploying a VoIP network is much less expensive than deploying a Private Branch Exchange (PBX).

2) **Rich media service and User Control Interface:** VoIP provides users with varied range of facilities like Instant Messaging, Status Control, Video Teleconferencing, Image and file exchange, and so on. It also comes with future possibilities like connectivity in mobile phones. VoIP providers allow users to dynamically change configuration settings through website links and rich GUIs.

3) **Phone and Service Portability:** Unlike the legacy PSTN, IP Phones are not confined to any physical address with a dedicated phone line attached to them. An IP Phone can use the same number virtually anywhere in the world as long as it is supported by adequate IP connectivity. It is a fully mobile device whose owner can travel everywhere irrespective of geographical boundaries. Along with the basic IP Phone setup, available telephony services like call features, voicemail access, call logs, security policies, etc. are also portable.

4) **Integration with Other Applications and Improved Productivity:** VoIP protocols (such as Session Initiation Protocol [SIP], H.323) run on the application layer. So, they can be merged with high-end applications like Instant Messengers, Chatting Applications, Mailing Systems, etc. VPN merged with VoIP can be used to set up entire office telephony, provided broadband connectivity is available. VoIP treats voice or video as data packets, so a user can also attach files along with voice or video. This proves tremendously useful in video-conferences.

5) **Location Independence:** The VoIP service area has no geographical limit. In other words, VoIP is not bound to the area code or country code of a specific location. It means any incoming call will be automatically routed to the VoIP phone number irrespective of its location, provided the user has already registered his/her location change with the VoIP server.

6) **Rich Features:** VoIP provides rich features like click-to-call that enable a user to initiate a call over a VoIP network by

simply clicking a URL on a web page. VoIP also provides many other facilities like Find-Me-Follow-Me (FMFM), selective call forwarding, personalized ring tones (or ring back tone), simultaneous rings on multiple phones, selective area or country code, call black list, call on hold etc.

7.2.3 VoIP Design Components

VoIP service network consists of the following design components:

1. Call Agent/SIP Client/SIP Server
2. Service Broker
3. Application Server
4. Media Server
5. Signalling Gateway
6. Trunking Gateway
7. Access Gateway
8. Access Concentrator
9. Bandwidth Manager
10. Edge Router
11. Subscriber Gateway
12. Bridge/Router
13. IP Phone/PBX

7.2.4 VoIP Security Issues

Like any other emerging IT technology, VoIP also comes with a large number of inherent and associated security risks and vulnerabilities that can affect the reliability of IT infrastructure of an organization. Therefore it is paramount for any organization looking to incorporate VoIP to have a total understanding of its risks and vulnerabilities. The risks can be broadly classified in three categories—inherited from IP, associated with VoIP, and specific to VoIP. Attacks on VoIP can also be of several different types, as described below:

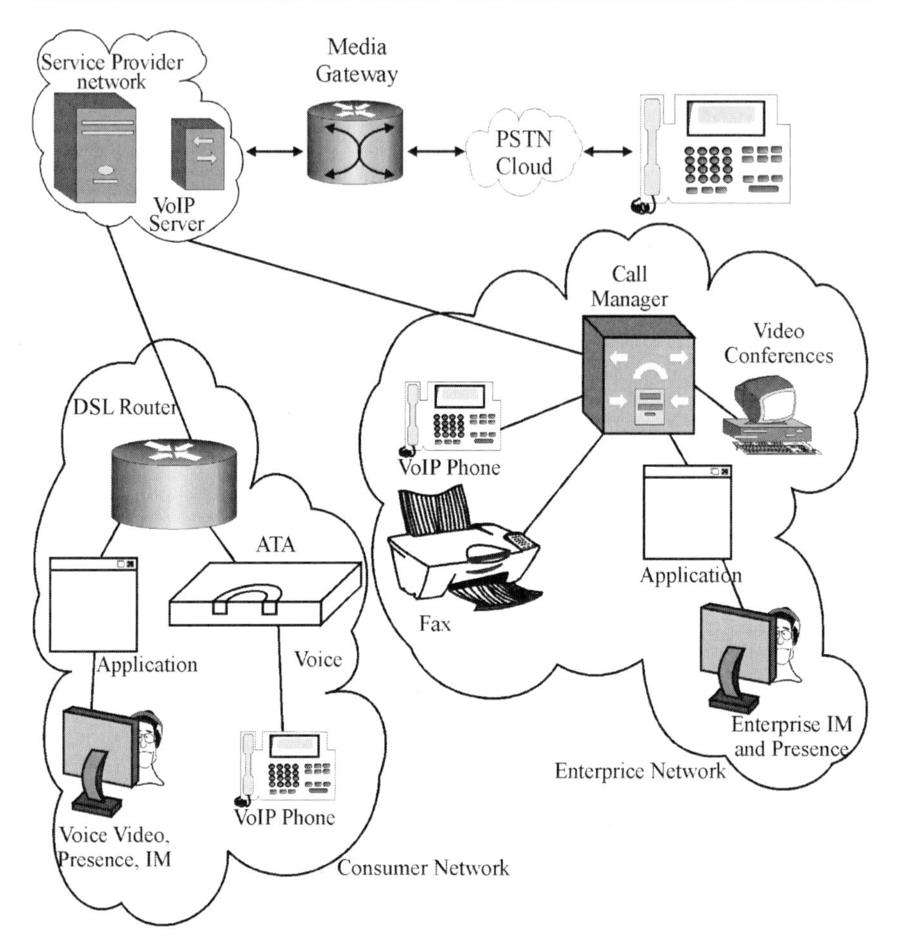

Figure 7.2: VoIP Architecture Diagram.

1) **Replay Attacks:** In this attack, a valid packet is captured and replayed into a network. The recipient device is tricked into reprocessing the data and re-responding, thereby generating even more traffic. With more packets from victim device it becomes easy to find more information leading to packer spoofing and masquerading or rather information to simply intervene into the target network.

2) **Packet Spoofing and Masquerading:** In packet spoofing, the origin address of a packet is spoofed or obscured to fool the victim device into believing that the packet is coming from a trusted source.

3) **Reconnaissance Attack:** By call walking and port scanning, an attacker first collects the necessary intelligence about the target network and then attempts to penetrate the network.
4) **Fuzzing:** In this attack, the attacker feeds the victim device semi-valid input and observes its reactions. Depending on the reaction, the attacker refines his/her input that results in more focused attacks.
5) **Denial of Service (DoS):** In DoS or Distributed DoS, the attacker deliberately sends a large amount of random data to one or more VoIP endpoints from a single source or from multiple sources. As a result, the system exhausts all its resources in processing the junk data and eventually denies service to legitimate users.
6) **CID Spoofing:** This is a type of masquerading attack based on manipulation of the Caller ID (CID). The attacker generally spoofs a CID and masquerades as a legitimate user in the network.
7) **Phone Impersonation:** It occurs due to the weak authentication mechanism in VoIP. In phone impersonation, an attacker hijacks passwords and impersonates as a legitimate user.
8) **Eavesdropping:** In between Real-time Transport Protocol (RTP) packet relay, when an attacker intervenes and actively steals packets flowing between legitimate users, it is known as a "VoIP eavesdrop attack".
9) **Call Hijacking and Redirection:** It is a special kind of active eavesdropping and man-in-the-middle type of attack where the eavesdropper hijacks calls between active users and redirects them to an unintended illegal recipient.

7.2.5 Security Measures for VoIP

There can be several remedial steps for avoiding security vulnerabilities associated with a VoIP network. The simplest ones are listed below:

1. **Encryption at the End Points:** With the assumption that endpoints are computationally powerful enough to handle

the encryption mechanism, there can be a possible solution for avoiding encryption bottlenecks at routers. The solution is to handle encryption and decryption only at the endpoints. It is not always feasible to protect packets at every hop of their lifetime by encrypting and decrypting them. Thus with advanced IP phones, encoding at end points accelerates the whole process.

2. **Secure Real Time Protocol (SRTP):** The Real-time Transport Protocol is commonly used to transmit real-time audio/video data in internet telephony applications. It is an insecure mechanism, susceptible to active eavesdropping, manipulation and replay of RTP data. Modified Real-time Transport Control Protocol data can even disrupt the processing of the RTP stream.

   In this situation, the Secure Real-time Protocol is a profile of the Real-time Transport Protocol (RTP) offering that not only confidentiality, but also message authentication and replay protection for the RTP as well as Real-time Transport Control Protocol (RTCP) traffic. The SRTP provides a security framework integrated with encryption algorithms like AES and key management techniques that achieve high throughput and low packet expansion.

3. **Key Management for SRTP—MIKEY:** The Multimedia Internet Keying (MIKEY) is designed to work with SRTP. MIKEY describes a real time key management scheme for multimedia data by the establishment of key materials with a two-way handshake. MIKEY comes with several key management schemes (e.g., the Diffie-Hellman key exchange). It provides better flexibility, casting and re-keying supports.

4. **Better Scheduling Schemes:** Although the introduction of encryption mechanisms and other security features reduces the scope of packet loss and eavesdropping, it also makes the system less speedy which may lead to DoS condition or starvation of urgent VoIP packets. A proper scheduling scheme should therefore be introduced (keeping quality of service (QoS) in mind) to enhance the crypto-engine's FIFO scheduler.

Recommended Readings and Web References

Dwivedi, Himanshu. 2005. Securing Storage: A Practical Guide to SAN and NAS Security; Addison-Wesley Professional; November 2005.

Park, Patrick. 2008. Voice over IP Security; Cisco Press; September 2008.

Ransome, James F. and John W. Rittinghouse. 2005. VoIP security; Elsevier Digital Press.

Reading Room SANS: An Overview of Storage Area Network from Security Perspective.

SecureLogix WebSite: Secure VoIP.

Tate, Jon, Fabiano Lucchese and Richard Moore. 2006. Introduction to Storage Area Networks; International Business Machines Corporation; July 2006.

A

Access Control List (ACL) 40
Advanced Encryption Standard (AES) 20
Alert Protocol 85
ASP.Net Authentication 119
  Forms 120
  None 120
  Passport 120
  Windows 120

B

Billion Laughs Attack 104
Black List checking 98
Buffer overflow 112
Bump in the Wire 67

C

Call Hijacking and Redirection 143
CAs 29
CBC (Cipher Block Chaining) 17
Change Cipher Spec Protocol 85
change_cipher_spec 84
ChangeCipher 88
CID Spoofing 143
Cisco ASA 5500 Series 39
Code Access Security (CAS) 116
Compression 60
Confidentiality 4, 23, 60
Configuration files 125
  ApplicationName.config 125
  Machine.config 125
  Web.config 125
CRC 61
Cross site scripting (XSS) 101
Cross-site request forgery (CSRF) 113

D

Data Encryption Algorithm (DEA)/ Data Encryption Standard (DES) 17
Data Integrity 24
Denial of Service (DoS), DoS and DDoS, DoS or Distributed DoS 9, 104, 143
Detective 12
Deterrent 12
Diffie-Hellman Key Exchange 26, 30, 75
Domain of Interpretation (DOI) 63
DTD 104

E

Eavesdropping 78, 143
ECB (Electronic Codebook) 17

Elliptic Curve Architecture and Cryptography 27
Email Compatibility 60
Encapsulating Security Payload (ESP) 62
External Threats 6

F

Fibre Channel Switches 135
Find-Me-Follow-Me (FMFM) 141
Form-based Authentication 131

H

H-IDS 48

G

Gateway 141
Gigabit Interface Converters 135

I

ICMP 65
ICMP Attack 10
IIS authentication 123
  Anonymous 125
  Basic 123
  Client Certificate Mapping 125
  Digest 124
  Integrated Windows Authentication 124
Internal Threats 5
Internet Key Exchange (IKE) 75
Internet Security Association and Key Management Protocol (ISAKMP) 62
Intrusion Detection and Prevention System (IDPS) 49
Intrusion Detection System 48
Intrusion Prevention System (IPS) 47

K

Kerberos 51, 52
Key Distribution Centre (KDC) 21, 52
Key Exchange 86, 87
Key Management for SRTP—MIKEY 144
key management 63

L

LDAP Injection 111
Log Injection 108
Logical Units (LUNs) 138

M

MachineKey Configuration 118
Man-in-the-middle 135, 136

N

N-IDS 48
Network-attached Storage (NAS) 134

O

OS-Integrated 64

P

Packet Filtering Router 34
Packet Spoofing and Masquerading 142
Personal Firewall 37
Pretty Good Privacy (PGP) 59
Preventive 12
Private Branch Exchange (PBX) 140
Public Key Infrastructure (PKI) 138
Public switched Telephone Network (PSTN) 139

R

Reconnaissance Attack 143
Replay Attacks 142
Router Implementation 66
RSA 25

S

SA Database (SADB) 68
Secure Real Time Protocol (SRTP) 144
Secure Socket Layer (SSL), Transport Layer Security (TLS), SSL/TLS 80

Security Association (SA) 67
Security Policy Database (SPD) 68
Security Triangle 3
Segmentation 61
Simple Network Management Protocol (SNMP) 137
SPI 68, 69
SQL Injection 95
SSL Record Protocol 83
Stateful Inspection Firewalls 35
Stateless Firewalls 35
Storage Area Network (SAN) 134

T

TCP SYN Flood 10
Ticket Granting Server 52
Ticket-to-get-ticket 52
Transport mode 71
Triple Data Encryption Algorithm (TDEA), Triple DES (3DES) 18
Tunnel mode 72
Type-safe SQL Parameters 99

V

Voice over Internet Protocol (VoIP) 139
VPN 39

W

White List checking 99
Windows CardSpace 117

X

X.509 29, 56
  Authentication Service 56
  certificates 87
  public-key certificates 56